THE CARUS METHEMATICAL MONOGRAPHS are an expression of the desire of Mrs. Mary Hegeler Carus, and of her son, Dr. Edward H. Carus, to contribute to the dissemination of mathematical knowledge by making accessible at nominal cost a series of expository presentations of the best thoughts and keenest researches in pure and applied mathematics. The publication of the first four of these monographs was made possible by a notable gift to the Mathematical Association of America by Mrs. Carus as sole trustee of the Edward C. Hegeler Trust Fund. The sales from these have resulted in the Carus Monograph Fund, and the Mathematical Association has used this as a revolving book fund to publish the succeeding monographs.

The expositions of mathematical subjects which the monographs contain are set forth in a manner comprehensible not only to teachers and students specializing in mathematics, but also to scientific workers in other fields, and especially to the wide circle of thoughtful people who, having a moderate acquaintance with elementary mathematics, wish to extend their knowledge without prolonged and critical study of the mathematical journals and treatises. The scope of this series includes also historical and biographical monographs.

The following monographs have been published:

No. 1. Calculus of Variations, G. A. BLISS

No. 2. Analytic Functions of a Complex Variable, by D. R. CURTISS

No. 3. Mathematical Statistics, by H. L. RIETZ

No. 4. Projective Geometry, by J. W. YOUNG

No. 5. A History of Mathematics in America before 1900, by D. E. SMITH and JEKUTHIEL GINSBURG (out of print)

No. 6. Fourier Series and Orthogonal Polynominals, by DUNHAM JACKSON

No. 7. Vectors and Matrices, by C. C. MacDUFFEE

No. 8. Rings and Ideals, by N. H. McCOY

RANDOM WALKS AND ELECTRIC NETWORKS

No. 9. The Theory of Algebraic Numbers, Second edition, by HARRY POLLARD and HAROLD G. DIAMOND

No. 10. The Arithmetic Theory of Quadratic Forms, by B. W. JONES

No. 11. Irrational Numbers, by IVAN NIVEN

No. 12. Statistical Independence in Probability, Analysis and Number Theory, by MARK KAC

No. 13. A Primer of Real Functions, Third edition, by RALPH P. BOAS, JR.

No. 14. Combinatorial Mathematics, by HERBERT JOHN RYSER

No. 15. Noncommutative Rings, by I. N. HERSTEIN

No. 16. Dedekind Sums, by HANS RADEMACHER and EMIL GROSSWALD

No. 17. The Schwarz Function and its Applications, by PHILIP J. DAVIS

No. 18. Celestial Mechanics, by HARRY POLLARD

No. 19. Field Theory and its Classical Problems, by CHARLES ROBERT HADLOCK

No. 20. The Generalized Riemann Integral, by ROBERT M. MCLEOD

No. 21. From Error-Correcting Codes through Sphere Packings to Simple Groups, by THOMAS M. THOMPSON.

No. 22. Random Walks and Electric Networks, by PETER G. DOYLE and J. LAURIE SNELL

THE

CARUS MATHEMATICAL MONOGRAPHS

Published by

THE MATHEMATICAL ASSOCIATION OF AMERICA

The Carus Mathematical Monographs

NUMBER TWENTY-TWO

RANDOM WALKS AND ELECTRIC NETWORKS

By

PETER G. DOYLE
J. LAURIE SNELL
Dartmouth College

Published and Distributed by
THE MATHEMATICAL ASSOCIATION OF AMERICA
Washington, D.C.

The Mathematical Association of America
1529 Eighteenth Street, NW
Washington, D.C. 20036

Printed in the United States of America.
Second printing, 1988.

The Library of Congress has cataloged the first printing of this title as follows:

Doyle, Peter G.
Random walks and electric networks / by Peter G. Doyle, J. Laurie Snell.—Washington, D.C.: Mathematical Association of America, c1984.

xiii, 159 p.: ill.; 20 cm.—(Carus Mathematical Monographs; no. 22)
Bibliography: p. 151–153.
Includes index.
ISBN 0-88385-024-9

1. Random walks (Mathematics) 2. Electric network topology. I. Snell, J. Laurie (James Laurie), 1925– . II. Title. III. Series.
QA274.73.D69 1984 519.2′82—dc19 84-61495

ISBN 0-88385-024-9 (No. 22)
ISBN 0-88385-000-1 (Complete Set)

PREFACE

Probability theory, like much of mathematics, is indebted to physics as a source of problems and intuition for solving these problems. Unfortunately, the level of abstraction of current mathematics often makes it difficult for anyone but an expert to appreciate this fact. In this book we will look at the interplay of physics and mathematics in terms of an example where the mathematics involved is at the college level. The example is the relation between elementary electric network theory and random walks.

Central to the book will be Pólya's beautiful theorem that a random walker on an infinite street network in d-dimensional space is bound to return to the starting point when $d = 2$, but has a positive probability of escaping to infinity without returning to the starting point when $d = 3$. Our goal will be to interpret this theorem as a statement about electric networks, and then to prove the theorem using techniques from classical electrical theory. The techniques referred to go back to Lord Rayleigh, who introduced them in connection with an investigation of musical instruments. The analog of Pólya's theorem in this connection is that wind instruments are possible in our three-dimensional world, but are not possible in Flatland (Abbott [**1**]).

The connection between random walks and electric networks has been recognized for some time (see Kakutani [**13**], Kemeny, Snell, and Knapp [**15**], and Kelly [**14**]). As for Rayleigh's method, the authors first learned it from Peter's father Bill Doyle, who used it to explain a mysterious comment in Feller ([**5**], p. 425, Problem 14). This comment suggested that a random walk in two dimensions remains recurrent when some of the streets are blocked, and while this is ticklish to prove probabilistically, it is an easy consequence of Rayleigh's method. The first person to apply Rayleigh's method to random walks seems to have been Nash-Williams [**26**]. Earlier, Royden [**32**] had applied Rayleigh's method to an equivalent problem. However, the true importance of Rayleigh's method for probability theory is only now becoming appreciated. See, for example, Griffeath and Liggett [**10**], Lyons [**22**], and Kesten [**17**].

Here's the plan of the book: In Part I we will restrict ourselves to the study of random walks on finite networks. Here we will establish the connection between the electrical concepts of current and voltage and corresponding descriptive quantities of random walks regarded as finite state Markov chains. In Part II we will consider random walks on infinite networks. Pólya's theorem will be proved using Rayleigh's method, and the proof will be compared with the classical proof using probabilistic methods. We will then discuss walks on more general infinite graphs, and use Rayleigh's method to derive certain extensions of Pólya's theorem. Certain of the results in Part II were obtained by Peter Doyle in work on his Ph.D. thesis.

To read this book, you should have a knowledge of the basic concepts of probability theory as well as a little electric network theory and linear algebra. An elementary introduction to finite Markov chains as presented by Kemeny, Snell, and Thompson [**16**] would be helpful.

The work of Snell was carried out while enjoying the hospitality of Churchill College and the Cambridge Statistical Laboratory supported by an NSF Faculty Development Fellowship. He thanks Professors Kendall and Whittle for making this such an enjoyable and rewarding visit. Peter Doyle thanks his father for teaching him how to think like a physicist. We both thank Peter Ney for assigning the problem in Feller that started all this, David Griffeath for suggesting the example to be used in our first proof that 3-dimensional random walk is recurrent (Part II Section 6.9), and Reese Prosser for keeping us going by his friendly and helpful hectoring. Special thanks are due Marie Slack, our secretary extraordinaire, for typing the original and the excessive number of revisions one is led to by computer formatting.

CONTENTS

PART I

Random Walks on Finite Networks

CHAPTER 1

RANDOM WALKS IN ONE DIMENSION

1.1 A Random Walk along Madison Avenue

A **random walk**, or **drunkard's walk**, was one of the first chance processes studied in probability; this chance process continues to play an important role in probability theory and its applications. An example of a random walk may be described as follows:

A man walks along a 5-block stretch of Madison Avenue. He starts at corner x and, with probability 1/2, walks one block to the right and, with probability 1/2, walks one block to the left; when he comes to the next corner he again randomly chooses his direction along Madison Avenue. He continues until he reaches corner 5, which is home, or corner 0, which is a bar. If he reaches either home or the bar, he stays there.

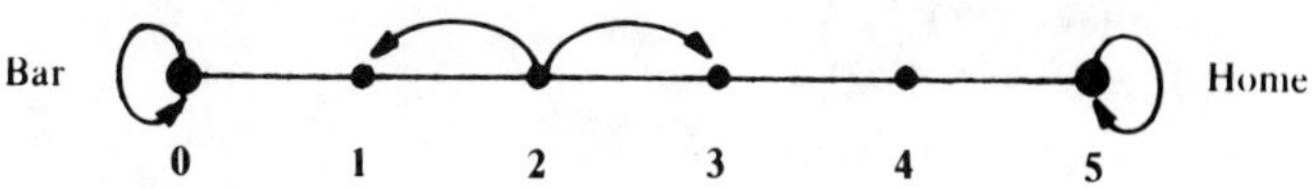

FIGURE 1.1

The problem we pose is to find the probability $p(x)$ that the man, starting at corner x, will reach home before reaching the bar. In looking at this problem, we will not be so much concerned with the particular form of the solution, which turns out to be $p(x) = x/5$, as with its general properties, which we will eventually describe by saying "$p(x)$ is the unique solution to a certain Dirichlet problem."

1.2 The Same Problem as a Penny Matching Game

In another form, the problem is posed in terms of the following game: Peter and Paul match pennies; they have a total of 5 pennies; on each match, Peter wins one penny from Paul with probability 1/2 and loses one with probability 1/2; they play until Peter's fortune reaches 0 (he is ruined) or reaches 5 (he wins all Paul's money). Now $p(x)$ is the probability that Peter wins if he starts with x pennies.

1.3 The Probability of Winning: Basic Properties

Consider a random walk on the integers $0, 1, 2, \ldots, N$. Let $p(x)$ be the probability, starting at x, of reaching N before 0. We regard $p(x)$ as a function defined on the points $x = 0, 1, 2, \ldots, N$. The function $p(x)$ has the following properties:

(a) $p(0) = 0$
(b) $p(N) = 1$
(c) $p(x) = (1/2)p(x - 1) + (1/2)p(x + 1)$
for $x = 1, 2, \ldots, N - 1$.

Properties (a) and (b) follow from our convention that 0 and N are traps; if the walker reaches one of these positions, he stops there; in the game interpretation, the game ends when one player has all of the pennies.

Property (c) states that, for an interior point, the probability $p(x)$ of reaching home from x is the average of the probabilities $p(x - 1)$ and $p(x + 1)$ of reaching home from the points that the walker may go to from x. We can derive (c) from the following basic fact about probability:

BASIC FACT. *Let E be any event, and F and G be events such that one and only one of the events F or G will occur. Then*

$$P(E) = P(F) \cdot P(E \text{ given } F) + P(G) \cdot P(E \text{ given } G).$$

In this case, let E be the event "the walker ends at the bar," F the event "the first step is to the left," and G the event "the first step is to the right." Then, if the walker starts at x, $P(E) = p(x)$, $P(F) = P(G) = 1/2$, $P(E \text{ given } F) = p(x - 1)$, $P(E \text{ given } G) = p(x + 1)$, and (c) follows.

1.4 An Electric Network Problem: The Same Problem?

Let's consider a second apparently very different problem. We connect equal resistors in series and put a unit voltage across the ends as in Figure 1.2.

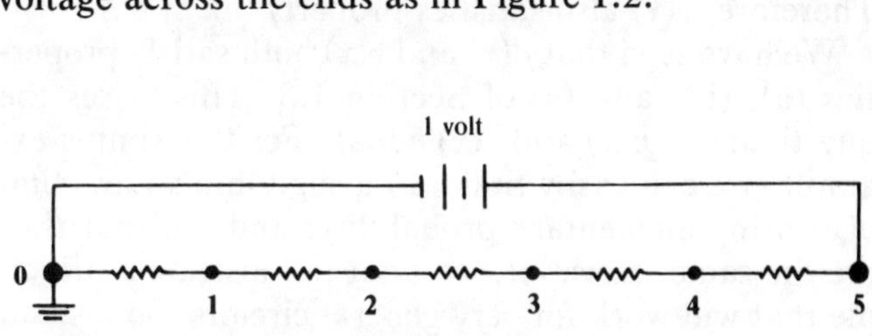

FIGURE 1.2

Voltages $v(x)$ will be established at the points $x = 0, 1, 2, 3, 4, 5$. We have grounded the point $x = 0$ so that $v(0) = 0$. We ask for the voltage $v(x)$ at the points x between the resistors. If we have N resistors, we make $v(0) = 0$ and $v(N) = 1$, so $v(x)$ satisfies properties (a) and (b) of Section 1.3. We now show that $v(x)$ also satisfies (c).

By Kirchhoff's Laws, the current flowing into x must be equal to the current flowing out. By Ohm's Law, if points x and y are connected by a resistance of magnitude R, then the current i_{xy} that flows from x to y is equal to

$$i_{xy} = \frac{v(x) - v(y)}{R}.$$

Thus for $x = 1, 2, \ldots, N - 1$

$$\frac{v(x-1) - v(x)}{R} + \frac{v(x+1) - v(x)}{R} = 0.$$

Multiplying through by R and solving for $v(x)$ gives

$$v(x) = \frac{v(x+1) + v(x-1)}{2}$$

$$\text{for} \quad x = 1, 2, \ldots, N - 1.$$

Therefore, $v(x)$ also satisfies property (c).

We have seen that $p(x)$ and $v(x)$ both satisfy properties (a), (b), and (c) of Section 1.3. This raises the question: are $p(x)$ and $v(x)$ equal? For this simple example, we can easily find $v(x)$ using Ohm's Law, find $p(x)$ using elementary probability, and see that they are the same. However, we want to illustrate a principle that will work for very general circuits. So instead we shall prove that these two functions are the same by showing that there is only one function that satisfies these properties, and we shall prove this by a method that will apply to more general situations than points connected together in a straight line.

Exercises

1. Referring to the random walk along Madison Avenue, let $X = p(1)$, $Y = p(2)$, $Z = p(3)$ and $W = p(4)$. Show that properties (a), (b), and (c) of Section 1.3 determine a set of four linear equations with variables X, Y, Z, and W; show that these equations have a unique solution. What does this say about $p(x)$ and $v(x)$ for this special case?

2. Assume that our walker has a tendency to drift in one direction: more specifically, assume that each step is to the right with probability p or to the left with probability $q = 1 - p$. Show that properties (a), (b), and (c) of Section 1.3 should be replaced by

   (a) $p(0) = 0$,
   (b) $p(N) = 1$,
   (c) $p(x) = q \cdot p(x - 1) + p \cdot p(x + 1)$.

3. In our electric network problem, assume that the resistors are not necessarily equal. Let R_x be the resistance between x and $x + 1$. Show that

$$v(x) = \frac{\dfrac{1}{R_{x-1}}}{\dfrac{1}{R_{x-1}} + \dfrac{1}{R_x}} v(x - 1) + \frac{\dfrac{1}{R_x}}{\dfrac{1}{R_{x-1}} + \dfrac{1}{R_x}} v(x + 1).$$

How should the resistors be chosen to correspond to the random walk of Exercise 2?

1.5 Harmonic Functions in One Dimension; the Uniqueness Principle

Let S be the set of points $S = \{0, 1, 2, \ldots, N\}$. We call the points of the set $D = \{1, 2, \ldots, N - 1\}$ the **interior points** of S and those of $B = \{0, N\}$ the **boundary points** of S. A function $f(x)$ defined on S is **harmonic** if, at points of D, it satisfies the averaging property

$$f(x) = \frac{f(x-1) + f(x+1)}{2}.$$

As we have seen, $p(x)$ and $v(x)$ are harmonic functions on S having the same values on the boundary: $p(0) = v(0) = 0$; $p(N) = v(N) = 1$. Thus both $p(x)$ and $v(x)$ solve the problem of finding a harmonic function having these boundary values. Now the problem of finding a harmonic function given its boundary values is called the **Dirichlet problem**, and the **Uniqueness Principle** for the Dirichlet problem asserts that there cannot be two different harmonic functions having the same boundary values. In particular, it follows that $p(x)$ and $v(x)$ are really the same function, and this is what we have been hoping to show. Thus the fact that $p(x) = v(x)$ is an aspect of a general fact about harmonic functions.

We will approach the Uniqueness Principle by way of the **Maximum Principle** for harmonic functions, which bears the same relation to the Uniqueness Principle as Rolle's Theorem does to the Mean Value Theorem of Calculus.

MAXIMUM PRINCIPLE. *A harmonic function $f(x)$ defined on S takes on its maximum value M and its minimum value m on the boundary.*

Proof. Let M be the largest value of f. Then if $f(x) = M$ for x in D, the same must be true for $f(x-1)$ and $f(x+1)$ since $f(x)$ is the average of these two values. If $x - 1$ is still an interior point, the same argument implies that $f(x-2) = M$; continuing in this way, we eventually conclude that $f(0) = M$. That same argument works for the minimum value m.

UNIQUENESS PRINCIPLE. *If $f(x)$ and $g(x)$ are harmonic functions on S such that $f(x) = g(x)$ on B, then $f(x) = g(x)$ for all x.*

Proof. Let $h(x) = f(x) - g(x)$. Then if x is any interior point,

$$\frac{h(x-1) + h(x+1)}{2} = \frac{f(x-1) + f(x+1)}{2} - \frac{g(x-1) + g(x+1)}{2}$$
$$= f(x) - g(x) = h(x)$$

and h is harmonic. But $h(x) = 0$ for x in B, and hence, by the Maximum Principle, the maximum and minimum values of h are 0. Thus $h(x) = 0$ for all x, and $f(x) = g(x)$ for all x.

Thus we finally prove that $p(x) = v(x)$; but what does $v(x)$ equal? The Uniqueness Principle shows us a way to find a concrete answer: just guess. For if we can find any harmonic function $f(x)$ having the right boundary values, the Uniqueness Principle guarantees that

$$p(x) = v(x) = f(x).$$

The simplest function to try for $f(x)$ would be a linear function; this leads to the solution $f(x) = x/N$. Note that $f(0) = 0$ and $f(N) = 1$ and

$$\frac{f(x-1)+f(x+1)}{2} = \frac{x-1+x+1}{2N} = \frac{x}{N} = f(x).$$

Therefore $f(x) = p(x) = v(x) = x/N$.

As another application of the Uniqueness Principle, we prove that our walker will eventually reach 0 or N. Choose a starting point x with $0 < x < N$. Let $h(x)$ be the probability that the walker never reaches the boundary $B = \{0, N\}$. Then

$$h(x) = (1/2)h(x+1) + (1/2)h(x-1)$$

and h is harmonic. Also $h(0) = h(N) = 0$; thus, by the Maximum Principle, $h(x) = 0$ for all x.

EXERCISES

1. Show that you can choose A and B so that the function $f(x) = A(q/p)^x + B$ satisfies the modified properties (a), (b) and (c) of Exercise 2 of Section 1.4. Does this show that $f(x) = p(x)$?

2. Let $m(x)$ be the expected number of steps, starting at x, required to reach 0 or N for the first time. It can be proven that $m(x)$ is finite. Show that $m(x)$ satisfies the conditions

   (a) $m(0) = 0$
   (b) $m(N) = 0$
   (c) $m(x) = \dfrac{1}{2}m(x+1) + \dfrac{1}{2}m(x-1) + 1.$

3. Show that conditions in Exercise 2 have a unique solution. Hint: show that if m and $\bar{m}$ are two solutions, then $f = m - \bar{m}$ is harmonic with $f(0) = f(N) = 0$ and hence $f(x) = 0$ for all x.

4. Show that you can choose A, B, and C such that $f(x) = A + Bx + Cx^2$ satisfies all the conditions of Exercise 2. Does this show that $f(x) = m(x)$ for this choice of A, B, and C?

5. Find the expected duration of the walk down Madison Avenue as a function of the walker's starting point (1, 2, 3, or 4).

1.6 The Solution as a Fair Game (Martingale)

Let us return to our interpretation of a random walk as Peter's fortune in a game of penny matching with Paul. On each match, Peter wins one penny with probability 1/2 and loses one penny with probability 1/2. Thus, when Peter has k pennies his expected fortune after the next play is

$$(1/2)\cdot(k - 1) + (1/2)\cdot(k + 1) = k,$$

so his expected fortune after the next play is equal to his present fortune. This says that he is playing a **fair game**; a chance process that can be interpreted as a player's fortune in a fair game is called a **martingale**.

Now assume that Peter and Paul have a total of N pennies. Let $p(x)$ be the probability that, when Peter has x pennies, he will end up with all N pennies. Then Peter's expected final fortune in this game is

$$(1 - p(x))\cdot 0 + p(x)\cdot N = p(x)\cdot N.$$

If we could be sure that a fair game remains fair to the end of the game, then we could conclude that Peter's expected final fortune is equal to his starting fortune x, i.e.,

$$x = p(x)\cdot N.$$

This would give $p(x) = x/N$ and we would have found the probability that Peter wins using the fact that a fair game remains fair to the end. Note that the time the game ends is a random time, namely, the time that the walk first reaches 0 or N for the first time. Thus the question is, is the fairness of a game preserved when we stop at a random time?

Unfortunately, this is not always the case. To begin with, if Peter somehow has knowledge of what the future holds in store for him, he can decide to quit when he gets to the end of a winning streak. But even if we restrict ourselves to stopping rules where the decision to stop or continue is independent of future events, fairness may not be preserved. For example, assume that Peter is allowed to go into debt and can play as long as he wants to. He starts with 0 pennies and decides to play until his fortune is 1 and then quit. We shall see that a random walk on the set of all integers, starting at 0, will reach the point 1 if we wait long enough. Hence, Peter will end up one penny ahead by this system of stopping.

However, there are certain conditions under which we can guarantee that a fair game remains fair when stopped at a random time. For our purposes, the following standard result of martingale theory will do:

MARTINGALE STOPPING THEOREM. *A fair game that is stopped at a random time will remain fair to the end of the game if it is assumed that there is a finite amount of money in the world and a player must stop if he wins all this money or goes into debt by this amount.*

This theorem would justify the above argument to obtain $p(x) = x/N$.

Let's step back and see how this martingale argument worked. We began with a harmonic function, the function $f(x) = x$, and interpreted it as the player's fortune in a fair game. We then considered the player's expected final fortune in this game. This was another harmonic function having the same boundary values and we appealed to the Martingale Stopping Theorem to argue that this function must be the same as the original function. This allowed us to write down an expression for the probability of winning, which was what we were looking for.

Lurking behind this argument is a general principle: If we are given boundary values of a function, we can come up with a harmonic function having these boundary values by assigning to each point the player's expected final fortune in a game where the player starts from the given point and carries out a random walk until he reaches a boundary point, where he receives the specified payoff. Furthermore, the Martingale Stopping Theorem allows us to conclude that there can be no other harmonic function with these boundary values. Thus martingale theory allows us to establish existence and uniqueness of solutions to a Dirichlet problem. All this isn't very exciting for the cases we've been considering, but the nice thing is that the same arguments carry through to the more general situations that we will be considering later on.

The study of martingales was originated by Lévy [**21**] and Ville [**36**]. Kakutani [**13**] showed the connection between random walks and harmonic functions. Doob [**4**] developed martingale stopping theorems and showed how to exploit the preservation of fairness to solve a wide variety of problems in probability theory. An informal discussion of martingales may be found in Snell [**34**].

Exercises

1. Consider a random walk with a drift; that is, there is a probability $p \neq 1/2$ of going one step to the right and a probability $q = 1 - p$ of going one step to the left. (See Exercise 2 of Section 1.4.) Let $w(x) = (q/p)^x$; show that, if you interpret $w(x)$ as your fortune when you are at x, the resulting game is fair. Then use the Martingale Stopping Theorem to argue that

$$w(x) = p(x)w(N) + (1 - p(x))w(0).$$

Solve for $p(x)$ to obtain

$$p(x) = \frac{\left(\dfrac{q}{p}\right)^x - 1}{\left(\dfrac{q}{p}\right)^N - 1}.$$

2. You are gambling against a professional gambler; you start with A dollars and the gambler with B dollars; you play a game in which you win one dollar with probability $p < 1/2$ and lose one dollar with probability $q = 1 - p$; play continues until you or the gambler runs out of money. Let R_A be the probability that you are ruined. Use the result of Exercise 1 to show that

$$R_A = \frac{1 - \left(\dfrac{p}{q}\right)^B}{1 - \left(\dfrac{p}{q}\right)^N}$$

with $N = A + B$. If you start with 20 dollars and the gambler with 50 dollars and $p = .45$, find the probability of being ruined.

3. The gambler realizes that the probability of ruining you is at least $1 - (p/q)^B$ (Why?). The gambler wants to make the probability at least .999. For this, $(p/q)^B$ should be at most .001. If the gambler offers you a game with $p = .499$, how large a stake should she have?

CHAPTER 2

RANDOM WALKS IN TWO DIMENSIONS

2.1 An Example

We turn now to the more complicated problem of a random walk on a two-dimensional array. In Figure 2.1 we illustrate such a walk. The large dots represent

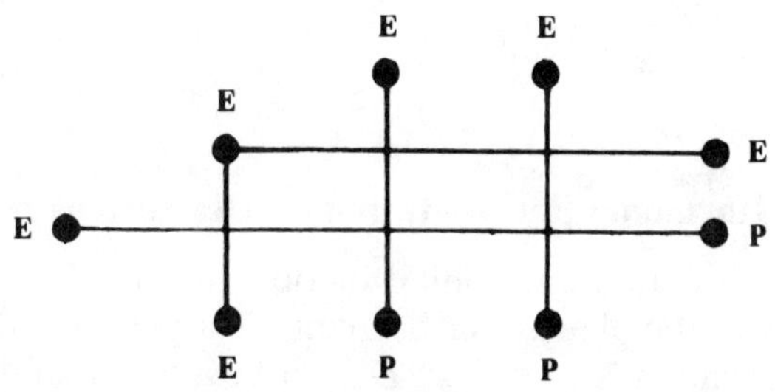

FIGURE 2.1

boundary points; those marked E indicate escape routes and those marked P are police. We wish to find the probability $p(x)$ that our walker, starting at an interior point x, will reach an escape route before he reaches a policeman. The walker moves from $x = (a, b)$ to each of the four neighboring points

$(a + 1, b)$, $(a - 1, b)$, $(a, b + 1)$, $(a, b - 1)$ with probability 1/4. If he reaches a boundary point, he remains at this point.

The corresponding voltage problem is shown in Figure 2.2. The boundary points P are grounded and points E are connected and fixed at one volt by a one-volt battery. We ask for the voltage $v(x)$ at the interior points.

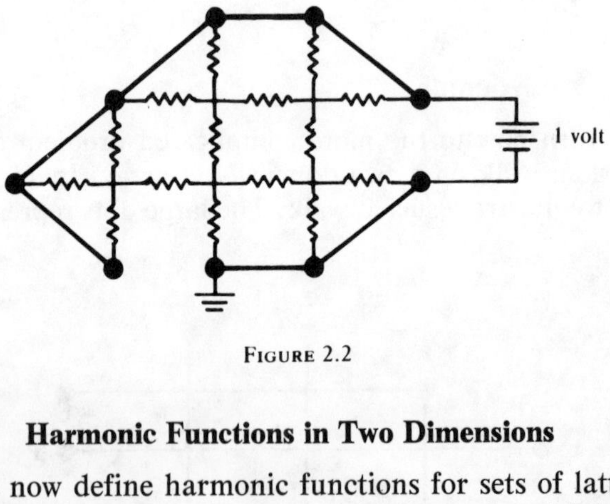

FIGURE 2.2

2.2 Harmonic Functions in Two Dimensions

We now define harmonic functions for sets of lattice points in the plane (a lattice point is a point with integer coordinates). Let $S = D \cup B$ be a finite set of lattice points such that (a) D and B have no points in common, (b) every point of D has its four neighboring points in S, and (c) every point of B has at least one of its four neighboring points in D. We assume further that S hangs together in a nice way, namely, that for any two points P and Q in S, there is a sequence of points P_j in D such that $P, P_1, \ldots, P_n, Q$ forms a path from P to Q. We call the points of D the **interior points** of S and the points of B the **boundary points** of S.

A function f defined on S is **harmonic** if, for points (a, b) in D, it has the averaging property

$$f(a, b) = \frac{f(a+1, b) + f(a-1, b) + f(a, b+1) + f(a, b-1)}{4}.$$

Note that there is no restriction on the values of f at the boundary points.

We would like to prove that $p(x) = v(x)$ as we did in the one-dimensional case. That $p(x)$ is harmonic follows again by considering all four possible first steps; that $v(x)$ is harmonic follows again by Kirchhoff's Laws since the current coming into $x = (a, b)$ is

$$\frac{v(a+1, b) - v(a, b)}{R} + \frac{v(a-1, b) - v(a, b)}{R} + \frac{v(a, b+1) - v(a, b)}{R} + \frac{v(a, b-1) - v(a, b)}{R} = 0.$$

Multiplying through by R and solving for $v(a, b)$ gives

$$v(a, b) = \frac{v(a+1, b) + v(a-1, b) + v(a, b+1) + v(a, b-1)}{4}.$$

Thus $p(x)$ and $v(x)$ are harmonic functions with the same boundary values. To show from this that they are the same, we must extend the Uniqueness Principle to two dimensions.

We first prove the Maximum Principle. If M is the maximum value of f and if $f(P) = M$ for P an interior point, then since $f(P)$ is the average of the values of f at its neighbors, these values must all equal M also. By working our way due south, say, repeating this argument at every step, we eventually reach a boundary point Q for which we can conclude that $f(Q) = M$. Thus a harmonic function always attains its maximum (or minimum) on the boundary; this is the Maximum Principle. The proof of the Uniqueness Principle goes through as before since again the difference of two harmonic functions is harmonic.

The fair game argument, using the Martingale Stopping Theorem, holds equally well and again gives an alternative proof of the existence and uniqueness to the solution of the Dirichlet problem.

Exercises

1. Show that if f and g are harmonic functions so is $h = a \cdot f + b \cdot g$ for constants a and b. This is called the **superposition principle**.

2. Let $B_1, B_2, \ldots, B_u$ be the boundary points for a region S. Let $e_j(a, b)$ be a function that is harmonic in S and has boundary value 1 at B_j and 0 at the other boundary points. Show that if arbitrary boundary values $v_1, v_2, \ldots, v_u$ are assigned, we can find the harmonic function v with these values from the solutions $e_1, e_2, \ldots, e_u$.

2.3 The Monte Carlo Solution

Finding the exact solution to a Dirichlet problem in two dimensions is not always a simple matter, so before taking on this problem, we will consider two methods for generating approximate solutions. In this section we will present a method using random walks. This method is known as a **Monte Carlo method**, since random walks are random, and gambling involves randomness, and there is a famous gambling casino in

Monte Carlo. In the next section, we will describe a much more effective method for finding approximate solutions, called the **method of relaxations**.

We have seen that the solution to the Dirichlet problem can be found by finding the value of a player's final winning in the following game: Starting at x the player carries out a random walk until reaching a boundary point. He is then paid an amount $f(y)$ if y is the boundary point first reached. Thus to find $f(x)$, we can start many random walks at x and find the average final winnings for these walks. By the law of averages (the law of large numbers in probability theory), the estimate that we obtain this way will approach the true expected final winning $f(x)$.

Here are some estimates obtained this way by starting 10,000 random walks from each of the interior points and, for each x, estimating $f(x)$ by the average winning of the random walkers who started at this point.

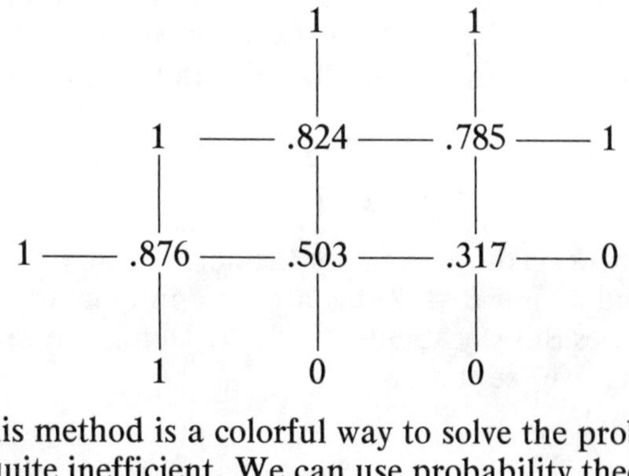

This method is a colorful way to solve the problem, but quite inefficient. We can use probability theory to estimate how inefficient it is. We consider the case with boundary values 1 or 0 as in our example. In this case, the expected final winning is just the probability that the walk ends up at a boundary point with value 1. For each point x, assume that we carry out n random walks; we regard each random walk to be an experiment and interpret the outcome of the ith experiment to be a "success" if the walker ends at a boundary point with a 1 and a "failure" otherwise. Let $p = p(x)$ be the unknown probability for success for a walker starting at x and $q = 1 - p$. How many walks should we carry out to get a reasonable estimate for p? We estimate p to be the fraction $\bar{p}$ of the walkers that end at a 1.

We are in the position of a pollster who wishes to estimate the proportion p of people in the country who favor candidate A over B. The pollster chooses a random sample of n people and estimates p as the proportion $\bar{p}$ of voters in his sample who favor A. (This is a gross oversimplification of what a pollster does—see Freedman et al. [8] for the real story.) To estimate the number n required, we can use the central limit theorem. This theorem states that, if S_n is the number of successes in n independent experiments, each having probability p for success, then for any $k > 0$

$$P\left(-k < \frac{S_n - np}{\sqrt{npq}} < k\right) \approx A(k)$$

where $A(k)$ is the area under the normal curve between $-k$ and k. For $k = 2$ this area is approximately .95; what does this say about $\bar{p} = S_n/n$? Doing a little rearranging, we see that

$$P\left(-2 < \frac{\bar{p} - p}{\sqrt{\frac{pq}{n}}} < 2\right) \approx .95$$

or

$$P\left(-2\sqrt{\frac{pq}{n}} < \bar{p} - p < 2\sqrt{\frac{pq}{n}}\right) \approx .95.$$

Since $\sqrt{pq} \leqslant 1/2$,

$$P\left(-\frac{1}{\sqrt{n}} < \bar{p} - p < \frac{1}{\sqrt{n}}\right) \gtrapprox .95.$$

Thus, if we choose $1/\sqrt{n} = .01$, or $n = 10{,}000$, there is a 95 percent chance that our estimate $\bar{p} = S_n/n$ will not be off by more than .01. This is a large number for rather modest accuracy; in our example we carried out 10,000 walks from each point and this required about 5 seconds on the Dartmouth computer. We shall see later, when we obtain an exact solution, that we did obtain the accuracy predicted.

Exercise

1. You play a game in which you start a random walk at the center in the grid shown in Figure 2.3. When the walk

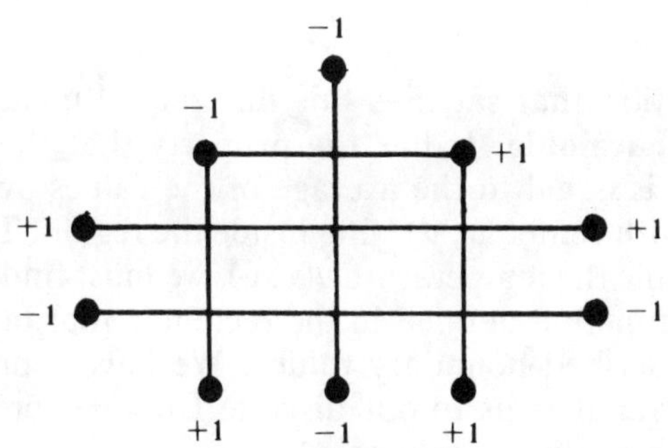

Figure 2.3

reaches the boundary, you receive a payment of $+1$ or -1 as indicated at the boundary points. You wish to simulate this game to see if it is a favorable game to play; how many simulations would you need to be reasonably certain of the value of this game to an accuracy of .01? Carry out such a simulation and see if you feel that it is a favorable game.

2.4 The Original Dirichlet Problem; the Method of Relaxations

The Dirichlet problem we have been studying is not the original Dirichlet problem, but a discrete version of it. The original Dirichlet problem concerns the distribution of temperature, say, in a continuous medium; the following is a representative example.

Suppose we have a thin sheet of metal gotten by cutting out a small square from the center of a large square. The inner boundary is kept at temperature 0 and the outer boundary is kept at temperature 1 as indicated in Figure 2.4. The problem is to find the temperature at points in the rectangle's interior. If $u(x, y)$ is the temperature at (x, y), then u satisfies Laplace's differential equation

$$u_{xx} + u_{yy} = 0.$$

A function that satisfies this differential equation is called **harmonic**. It has the property that the value $u(x, y)$ is equal to the average of the values over any circle with center (x, y) lying inside the region. Thus to determine the temperature $u(x, y)$, we must find a harmonic function defined in the rectangle that takes on the prescribed boundary values. We have a problem entirely analogous to our discrete Dirichlet problem, but with continuous domain.

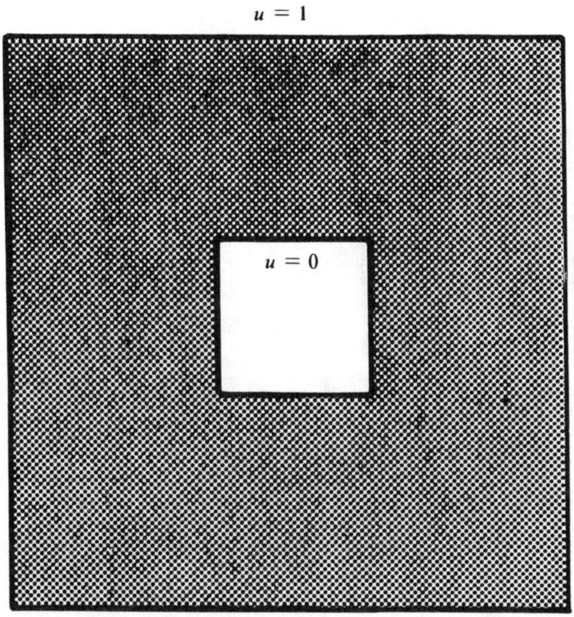

FIGURE 2.4

The **method of relaxations** was introduced as a way to get approximate solutions to the original Dirichlet problem. This method is actually more closely connected to the discrete Dirichlet problem than to the continuous problem. Why? Because, faced with the continuous problem just described, no physicist will hesitate to replace it with an analogous discrete problem, approximating the continuous medium by an array of lattice points such as that depicted in Figure 2.5, and searching for a function that is harmonic in our discrete sense and that takes on the appropriate

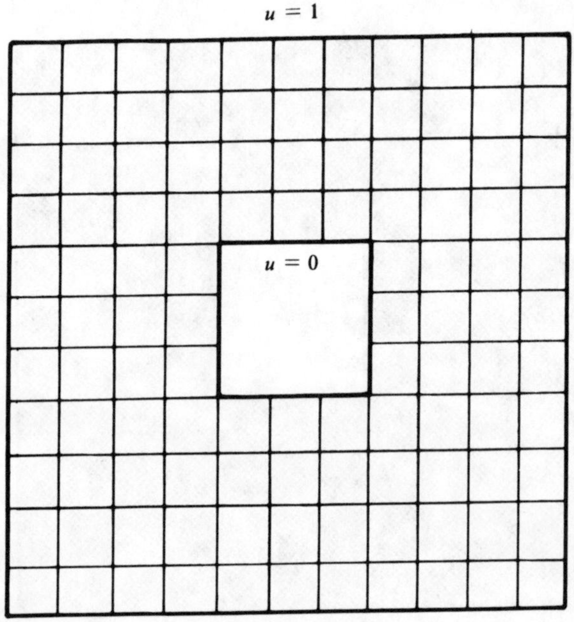

FIGURE 2.5

boundary values. It is this approximating discrete problem to which the method of relaxations applies.

Here's how the method goes. Recall that we are looking for a function that has specified boundary values, for which the value at any interior point is the average of the values at its neighbors. Begin with any function having the specified boundary values, pick an interior point, and see what is happening there. In general, the value of the function at the point we are looking at will not be equal to the average of the values at its neighbors. So adjust the value of the function to be equal to the average of the values at its neighbors. Now run through the rest of the interior points, repeating this process. When you have adjusted the values at all of the interior points, the function that results will not be harmonic, because most of the time after adjusting the value at a point to be the average value at its neighbors, we afterwards came along and adjusted the values at one or more of those neighbors, thus destroying the harmony. However, the function that results after running through all the interior points, if not harmonic, is more nearly harmonic than the function we started with; if we keep repeating this averaging process, running through all of the interior points again and again, the function will approximate more and more closely the solution to our Dirichlet problem.

We do not yet have the tools to prove that this method works for a general initial guess; this will have to wait until Chapter 3 (see Exercise 2 of Section 3.5). We will start with a special choice of initial values for which we can prove that the method works (see Exercise 2 of this section).

We start with all interior points 0 and keep the boundary points fixed.

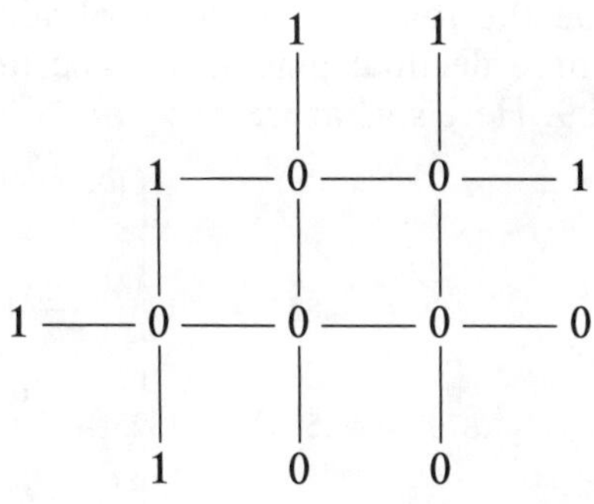

After one iteration we have:

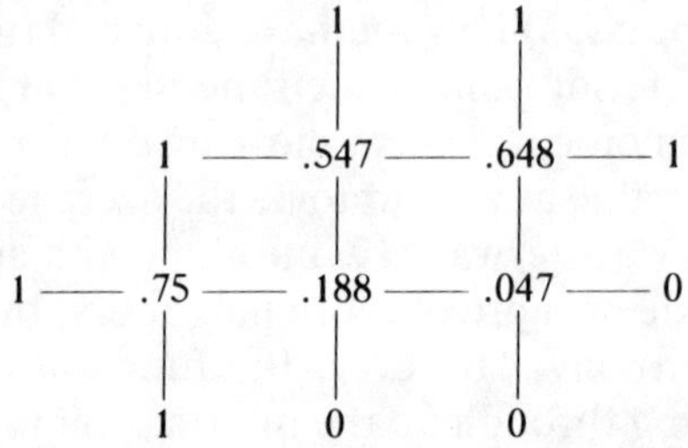

Note that we go from left to right moving up each column replacing each value by the average of the four neighboring values. The computations for this first iteration are

$$.75 = (1/4)(1 + 1 + 1 + 0)$$
$$.1875 = (1/4)(.75 + 0 + 0 + 0)$$
$$.5469 = (1/4)(.1875 + 1 + 1 + 0)$$
$$.0469 = (1/4)(.1875 + 0 + 0 + 0)$$
$$.64844 = (1/4)(.0469 + .5769 + 1 + 1)$$

We have printed the results to three decimal places. We continue the iteration until we obtain the same results to three decimal places. This occurs at iterations 8 and 9. Here's what we get:

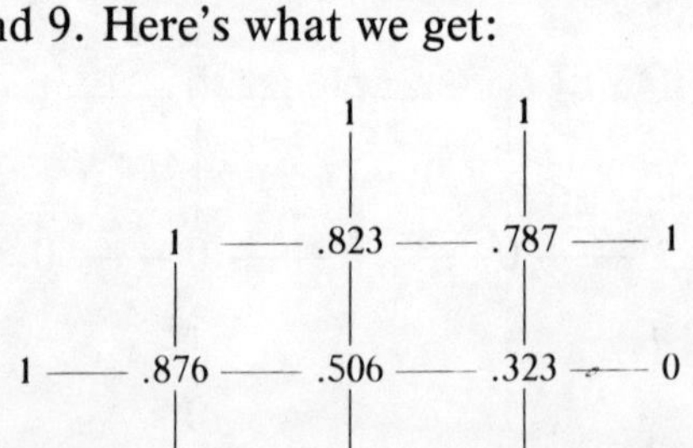

We see that we obtain the same result to three places after only nine iterations and this took only a fraction of a second of computing time. We shall see that these results are correct to three place accuracy. Our Monte Carlo method took several seconds of computing time and did not even give three place accuracy.

The classical reference for the method of relaxations as a means of finding approximate solutions to continuous problems is Courant, Friedrichs, and Lewy [**3**]. For more information on the relationship between the original Dirichlet problem and the discrete analog, see Hersh and Griego [**11**].

EXERCISES

1. Apply the method of relaxations to the discrete problem illustrated in Figure 2.5.

2. Consider the method of relaxations started with an initial guess with the property that the value at each point is $\leqslant$ the average of the values at the neighbors of this point. Show that the successive values at a point u are monotone increasing with a limit $f(u)$ and that these limits provide a solution to the Dirichlet problem.

2.5 Solution by Solving Linear Equations

In this section we will show how to find an exact solution to a two-dimensional Dirichlet problem by solving a system of linear equations. As usual, we will illustrate the method in the case of the example introduced in Section 2.1. This example is shown again in Figure 2.6; the interior points have been labelled a, b, c, d, and e. By our averaging property, we have

$$x_a = \frac{x_b + x_d + 2}{4}$$

$$x_b = \frac{x_a + x_e + 2}{4}$$

$$x_c = \frac{x_d + 3}{4}$$

$$x_d = \frac{x_a + x_c + x_e}{4}$$

$$x_e = \frac{x_b + x_d}{4}.$$

We can rewrite these equations in matrix form as

$$\begin{pmatrix} 1 & -1/4 & 0 & -1/4 & 0 \\ -1/4 & 1 & 0 & 0 & -1/4 \\ 0 & 0 & 1 & -1/4 & 0 \\ -1/4 & 0 & -1/4 & 1 & -1/4 \\ 0 & -1/4 & 0 & -1/4 & 1 \end{pmatrix} \begin{pmatrix} x_a \\ x_b \\ x_c \\ x_d \\ x_e \end{pmatrix} = \begin{pmatrix} 1/2 \\ 1/2 \\ 3/4 \\ 0 \\ 0 \end{pmatrix}$$

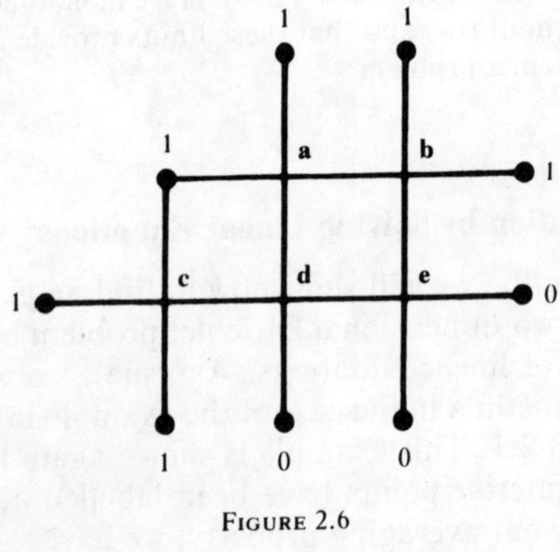

FIGURE 2.6

We can write this in symbols as

$$\mathbf{A}\mathbf{x} = \mathbf{u}.$$

Since we know there is a unique solution, **A** must have an inverse and

$$\mathbf{x} = \mathbf{A}^{-1}\mathbf{u}.$$

Carrying out this calculation we find

$$\text{Calculated } \mathbf{x} = \begin{pmatrix} .823 \\ .787 \\ .876 \\ .506 \\ .323 \end{pmatrix}$$

Here, for comparison, are the approximate solutions found earlier:

$$\text{Monte Carlo } \mathbf{x} = \begin{pmatrix} .824 \\ .785 \\ .876 \\ .503 \\ .317 \end{pmatrix} \quad \text{Relaxed } \mathbf{x} = \begin{pmatrix} .823 \\ .787 \\ .876 \\ .506 \\ .323 \end{pmatrix}$$

We see that our Monte Carlo approximations were fairly good in that no error of the simulation is greater than .01, and our relaxed approximations were very good indeed, in that the error does not show up at all.

Exercises

1. Consider a random walker on the graph of Figure 2.7. Find the probability of reaching the point with a 1 before any of the points with 0's for each starting point a, b, c, d.

2. Solve the discrete Dirichlet problem for the graph shown in Figure 2.8. The interior points are a, b, c, d. (Hint: See Exercise 2 of Section 2.2.)

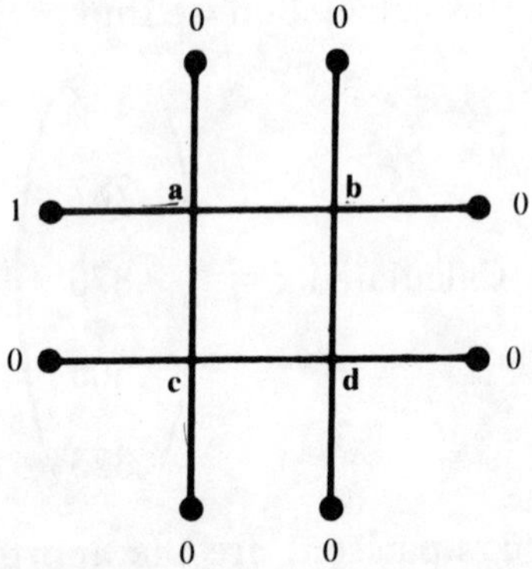

FIGURE 2.7

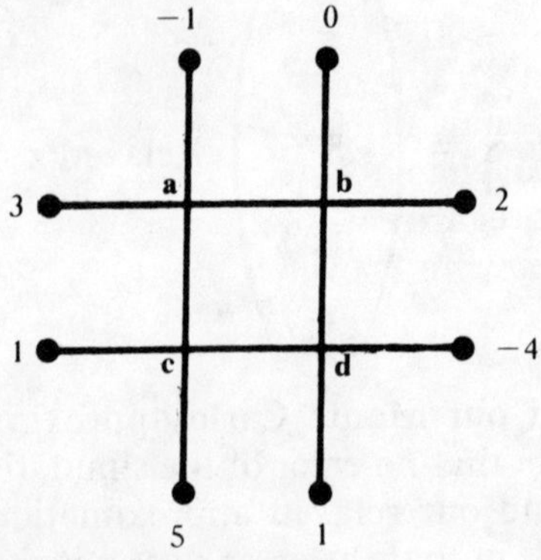

FIGURE 2.8

3. Find the exact value, for each possible starting point, for the game described in Exercise 1 of Section 2.3. Is the game favorable starting in the center?

2.6 Solution by the Method of Markov Chains

In this section, we describe how the Dirichlet problem can be solved by the method of Markov chains. This method may be viewed as a more sophisticated version of the method of linear equations.

A **finite Markov chain** is a special type of chance process that may be described informally as follows: we have a set $S = \{s_1, s_2, \ldots, s_r\}$ of **states** and a chance process that moves around through these states. When the process is in state s_i, it moves with probability P_{ij} to the state s_j. The **transition probabilities** P_{ij} are represented by an r-by-r matrix $\mathbf{P}$ called the **transition matrix.** To specify the chance process completely we must give, in addition to the transition matrix, a method for starting the process. We do this by specifying a specific state in which the process starts.

According to Kemeny, Snell, and Thompson [**16**], in the Land of Oz, there are three kinds of weather: rain, nice, and snow. There are never two nice days in a row. When it rains or snows, half the time it is the same the next day. If the weather changes, the chances are equal for a change to each of the other two types of weather. We regard the weather in the Land of Oz as a Markov chain with transition matrix:

$$
\begin{array}{c} \\ \mathbf{P} = \end{array}
\begin{array}{c c}
 & \begin{array}{ccc} R & N & S \end{array} \\
\begin{array}{c} R \\ N \\ S \end{array} &
\begin{pmatrix} 1/2 & 1/4 & 1/4 \\ 1/2 & 0 & 1/2 \\ 1/4 & 1/4 & 1/2 \end{pmatrix}
\end{array}.
$$

When we start in a particular state, it is natural to ask for the probability that the process is in each of the possible states after a specific number of steps. In the study of Markov chains, it is shown that this information is provided by the powers of the transition matrix. Specifically, if $\mathbf{P}^n$ is the matrix $\mathbf{P}$ raised to the nth power, the entries P_{ij}^n represent the probability that the chain, started in state s_i, will, after n steps, be in state s_j. For example, the fourth power of the transition matrix $\mathbf{P}$ for the weather in the Land of Oz is

$$\mathbf{P}^4 = \begin{array}{c} \\ R \\ N \\ S \end{array}\begin{array}{c} \begin{array}{ccc} R & N & S \end{array} \\ \begin{pmatrix} .402 & .199 & .398 \\ .398 & .203 & .398 \\ .398 & .199 & .402 \end{pmatrix} \end{array}.$$

Thus, if it is raining today in the Land of Oz, the probability that the weather will be nice four days from now is .199. Note that the probability of a particular type of weather four days from today is essentially independent of the type of weather today. This Markov chain is an example of a type of chain called a regular chain. A Markov chain is a **regular** chain if some power of the transition matrix has no zeros. In the study of regular Markov chains, it is shown that the probability of being in a state after a large number of steps is independent of the starting state.

As a second example, we consider a random walk in one dimension. Let us assume that the walk is stopped when it reaches either state 0 or 4. (We could use 5 instead of 4, as before, but we want to keep the matrices small.) We can regard this random walk as a Markov chain with states 0, 1, 2, 3, 4 and transition matrix given by

$$
\mathbf{P} = \begin{array}{c} \\ 0 \\ 1 \\ 2 \\ 3 \\ 4 \end{array}
\begin{array}{c} \begin{array}{ccccc} 0 & 1 & 2 & 3 & 4 \end{array} \\
\begin{pmatrix} 1 & 0 & 0 & 0 & 0 \\ 1/2 & 0 & 1/2 & 0 & 0 \\ 0 & 1/2 & 0 & 1/2 & 0 \\ 0 & 0 & 1/2 & 0 & 1/2 \\ 0 & 0 & 0 & 0 & 1 \end{pmatrix} \end{array}.
$$

The states 0 and 4 are **traps** or **absorbing states**. These are states that, once entered, cannot be left. A Markov chain is called **absorbing** if it has at least one absorbing state and if, from any state, it is possible (not necessarily in one step) to reach at least one absorbing state. Our Markov chain has this property and so is an absorbing Markov chain. The states of an absorbing chain that are not traps are called **non-absorbing**.

When an absorbing Markov chain is started in a non-absorbing state, it will eventually end up in an absorbing state. For non-absorbing state s_i and absorbing state s_j, we denote by B_{ij} the probability that the chain starting in s_i will end up in state s_j. We denote by $\mathbf{B}$ the matrix with entries B_{ij}. This matrix will have as many rows as non-absorbing states and as many columns as there are absorbing states. For our random walk example, the entries $B_{x,4}$ will give the probability that our random walker, starting at x, will reach 4 before reaching 0. Thus, if we can find the matrix $\mathbf{B}$ by Markov chain techniques, we will have a way to solve the Dirichlet problem.

We shall show, in fact, that the Dirichlet problem has a natural generalization in the context of absorbing Markov chains and can be solved by Markov chain methods.

Assume now that $\mathbf{P}$ is an absorbing Markov chain and that there are u absorbing states and v non-absorbing states. We reorder the states so that the absorbing states come first and the non-absorbing states come last. Then our transition matrix has the **canonical form**:

$$\mathbf{P} = \left(\begin{array}{c|c} \mathbf{I} & \mathbf{0} \\ \hline \mathbf{R} & \mathbf{Q} \end{array}\right).$$

Here $\mathbf{I}$ is a u-by-u identity matrix; $\mathbf{0}$ is a matrix of dimension u-by-v with all entries 0.

For our random walk example this canonical form is:

$$\mathbf{P} = \begin{array}{c} \\ 0 \\ 4 \\ 1 \\ 2 \\ 3 \end{array} \begin{array}{c} \begin{array}{ccccc} 0 & 4 & 1 & 2 & 3 \end{array} \\ \left(\begin{array}{cc|ccc} 1 & 0 & 0 & 0 & 0 \\ 0 & 1 & 0 & 0 & 0 \\ \hline 1/2 & 0 & 0 & 1/2 & 0 \\ 0 & 0 & 1/2 & 0 & 1/2 \\ 0 & 1/2 & 0 & 1/2 & 0 \end{array}\right) \end{array}$$

The matrix $\mathbf{N} = (\mathbf{I} - \mathbf{Q})^{-1}$ is called the **fundamental matrix** for the absorbing chain $\mathbf{P}$. The entries N_{ij} of this matrix have the following probabilistic interpretation: N_{ij} is the expected number of times that the chain will be in state s_j before absorption when it is started in s_i. (To see why this is true, think of how $(\mathbf{I} - \mathbf{Q})^{-1}$ would look if it were written as a geometric series.) Let $\mathbf{1}$ be a column vector of all 1's. Then the vector $\mathbf{t} = \mathbf{N1}$

gives the expected number of steps before absorption for each starting state.

The absorption probabilities **B** are obtained from **N** by the matrix formula

$$\mathbf{B} = (\mathbf{I} - \mathbf{Q})^{-1}\mathbf{R}.$$

This simply says that to get the probability of ending up at a given absorbing state, we add up the probabilities of going there from all the non-absorbing states, weighted by the number of times we expect to be in those (non-absorbing) states.

For our random walk example

$$\mathbf{Q} = \begin{pmatrix} 0 & 1/2 & 0 \\ 1/2 & 0 & 1/2 \\ 0 & 1/2 & 0 \end{pmatrix}$$

$$\mathbf{I} - \mathbf{Q} = \begin{pmatrix} 1 & -1/2 & 0 \\ -1/2 & 1 & -1/2 \\ 0 & -1/2 & 1 \end{pmatrix}$$

$$\mathbf{N} = (\mathbf{I} - \mathbf{Q})^{-1} = \begin{matrix} & \begin{matrix} 1 & \;\;2\;\; & 3 \end{matrix} \\ \begin{matrix} 1 \\ 2 \\ 3 \end{matrix} & \begin{pmatrix} 3/2 & 1 & 1/2 \\ 1 & 2 & 1 \\ 1/2 & 1 & 3/2 \end{pmatrix} \end{matrix}$$

$$\mathbf{t} = \mathbf{N1} = \begin{pmatrix} 3/2 & 1 & 1/2 \\ 1 & 2 & 1 \\ 1/2 & 1 & 3/2 \end{pmatrix} \begin{pmatrix} 1 \\ 1 \\ 1 \end{pmatrix} = \begin{pmatrix} 3 \\ 4 \\ 3 \end{pmatrix}$$

$$\mathbf{B} = \mathbf{NR} = \begin{pmatrix} 3/2 & 1 & 1/2 \\ 1 & 2 & 1 \\ 1/2 & 1 & 3/2 \end{pmatrix} \begin{pmatrix} 1/2 & 0 \\ 0 & 0 \\ 0 & 1/2 \end{pmatrix}$$

$$= \begin{matrix} \\ 1 \\ 2 \\ 3 \end{matrix} \begin{matrix} \begin{matrix} 0 & \quad 4 \end{matrix} \\ \begin{pmatrix} 3/4 & 1/4 \\ 1/2 & 1/2 \\ 1/4 & 3/4 \end{pmatrix} \end{matrix}$$

Thus, starting in state 3, the probability is 3/4 of reaching 4 before 0; this is in agreement with our previous results. From **t** we see that the expected duration of the game, when we start in state 2, is 4.

For an absorbing chain **P**, the *n*th power $\mathbf{P}^n$ of the transition probabilities will approach a matrix $\mathbf{P}^\infty$ of the form

$$\mathbf{P}^\infty = \left(\begin{array}{c|c} \mathbf{I} & \mathbf{0} \\ \hline \mathbf{B} & \mathbf{0} \end{array}\right).$$

We now give our Markov chain version of the Dirichlet problem. We interpret the absorbing states as boundary states and the non-absorbing states as interior states. Let B be the set of boundary states and D the set of interior states. Let f be a function with domain the state space of a Markov chain **P** such that for i in D

$$f(i) = \sum_j P_{ij} f(j).$$

Then f is a **harmonic function** for P. Now f again has an averaging property and extends our previous definition. If we represent f as a column vector $\mathbf{f}$, f is harmonic if and only if

$$\mathbf{Pf} = \mathbf{f}.$$

This implies that

$$\mathbf{P}^2\mathbf{f} = \mathbf{P} \cdot \mathbf{Pf} = \mathbf{Pf} = \mathbf{f}$$

and in general

$$\mathbf{P}^n\mathbf{f} = \mathbf{f}.$$

Let us write the vector $\mathbf{f}$ as

$$\mathbf{f} = \begin{pmatrix} \mathbf{f}_B \\ \mathbf{f}_D \end{pmatrix}$$

where $\mathbf{f}_B$ represents the values of $\mathbf{f}$ on the boundary and $\mathbf{f}_D$ values on the interior. Then we have

$$\begin{pmatrix} \mathbf{f}_B \\ \mathbf{f}_D \end{pmatrix} = \left(\begin{array}{c|c} \mathbf{I} & \mathbf{0} \\ \hline \mathbf{B} & \mathbf{0} \end{array}\right) \begin{pmatrix} \mathbf{f}_B \\ \mathbf{f}_D \end{pmatrix}$$

and

$$\mathbf{f}_D = \mathbf{Bf}_B.$$

We again see that the values of a harmonic function are determined by the values of the function at the boundary points.

Since the entries B_{ij} of $\mathbf{B}$ represent the probability, starting in i, that the process ends at j, our last equation states that if you play a game in which your fortune is f_j when you are in state j, then your expected final fortune is equal to your initial fortune; that is, fairness is preserved. As remarked above, from Markov chain theory $\mathbf{B} = \mathbf{NR}$ where $\mathbf{N} = (\mathbf{I} - \mathbf{Q})^{-1}$. Thus

$$\mathbf{f}_D = (\mathbf{I} - \mathbf{Q})^{-1}\mathbf{Rf}_B.$$

(To make the correspondence between this solution and the solution of Section 2.5, put $\mathbf{A} = \mathbf{I} - \mathbf{Q}$ and $\mathbf{u} = \mathbf{R}\mathbf{f}_B$).

A general discussion of absorbing Markov chains may be found in Kemeny, Snell, and Thompson [**16**].

Exercise

1. Consider the game played on the grid in Figure 2.9. You start at an interior point and move randomly until a boundary point is reached and obtain the payment indicated at this point. Using Markov chain methods find, for each starting state, the expected value of the game. Find also the expected duration of the game.

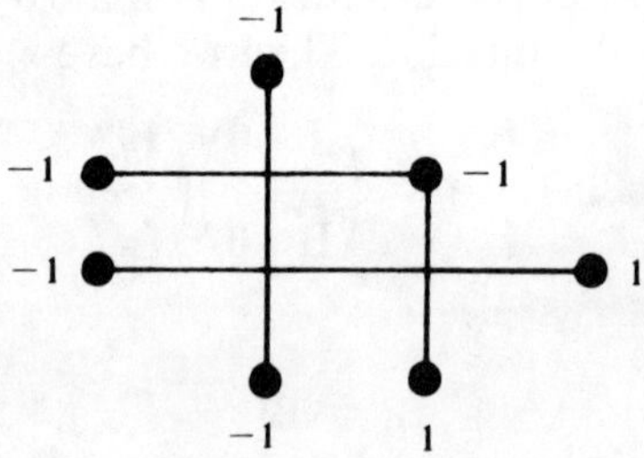

FIGURE 2.9

CHAPTER 3

RANDOM WALKS ON MORE GENERAL NETWORKS

3.1 General Resistor Networks and Reversible Markov Chains

Our networks so far have been very special networks with unit resistors. We will now introduce general resistor networks, and consider what it means to carry out a random walk on such a network.

A **graph** is a finite collection of **points** (also called **vertices** or **nodes**) with certain pairs of points connected by **edges** (also called **branches**). The graph is **connected** if it is possible to go between any two points by moving along the edges.

FIGURE 3.1

We assume that G is a connected graph and assign to each edge xy a resistance R_{xy}; an example is shown in Figure 3.2. The **conductance** of an edge xy is $C_{xy} = 1/R_{xy}$; the conductances for our example are shown in Figure 3.3.

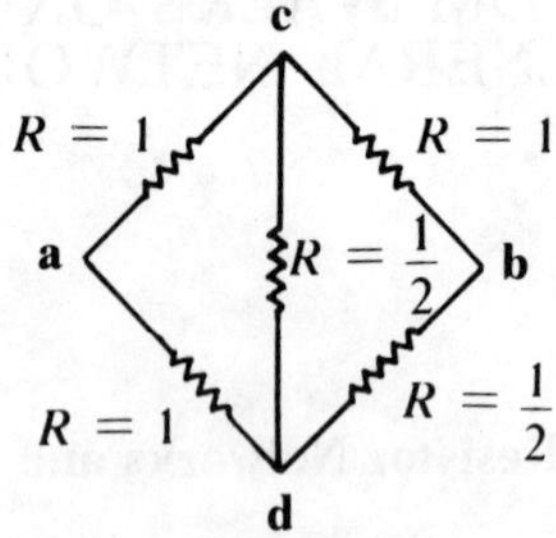

Resistances

FIGURE 3.2

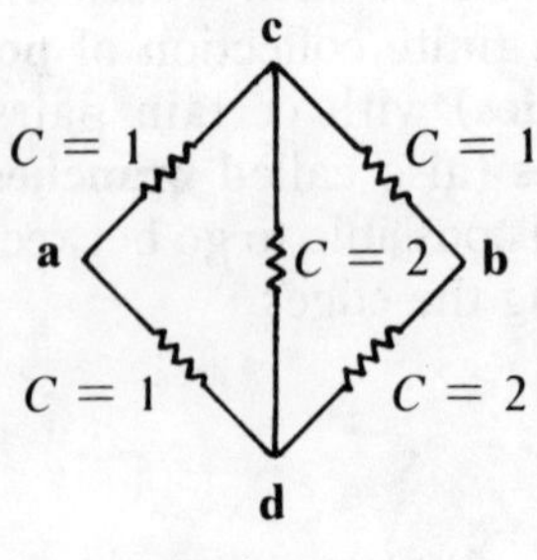

Conductances

FIGURE 3.3

We define a **random walk** on G to be a Markov chain with transition matrix **P** given by

$$P_{xy} = \frac{C_{xy}}{C_x}$$

with $C_x = \Sigma_y C_{xy}$. For our example, $C_a = 2$, $C_b = 3$, $C_c = 4$ and $C_d = 5$, and the transition matrix $\mathbf{P}$ for the associated random walk is

$$\mathbf{P} = \begin{array}{c} \\ a \\ b \\ c \\ d \end{array} \begin{array}{c} \begin{array}{cccc} a & b & c & d \end{array} \\ \begin{pmatrix} 0 & 0 & 1/2 & 1/2 \\ 0 & 0 & 1/3 & 2/3 \\ 1/4 & 1/4 & 0 & 1/2 \\ 1/5 & 2/5 & 2/5 & 0 \end{pmatrix} \end{array}.$$

Its graphical representation is shown in Figure 3.4.

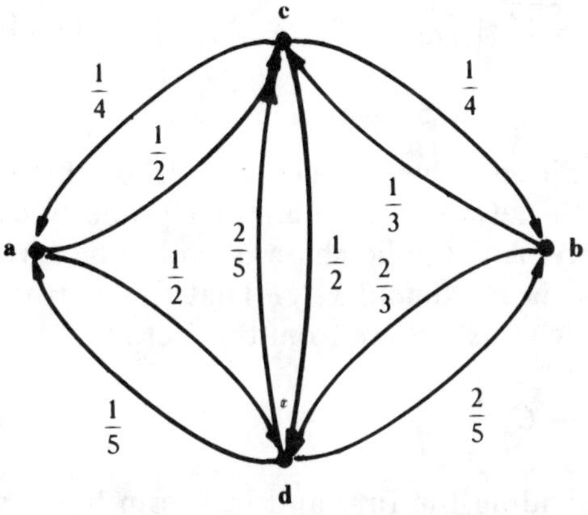

FIGURE 3.4

Since the graph is connected, it is possible for the walker to go between any two states. A Markov chain with this property is called an **ergodic Markov chain**. Regular chains, which were introduced in Section 2.6, are always ergodic, but ergodic chains are not always regular (see Exercise 1).

For an ergodic chain, there is a unique probability vector **w** that is a fixed vector for **P**, i.e., **wP** = **w**. The component w_j of **w** represents the proportion of times, in the long run, that the walker will be in state j. For random walks determined by electric networks, the fixed vector is given by $w_x = C_x/C$ where $C = \Sigma_x C_x$. (You are asked to prove this in Exercise 2.) For our example $C_a = 2$, $C_b = 3$, $C_c = 4$, $C_d = 5$ and $C = 14$. Thus **w** = (2/14, 3/14, 4/14, 5/14). We can check that **w** is a fixed vector by noting that

$$\begin{pmatrix}\frac{2}{14} & \frac{3}{14} & \frac{4}{14} & \frac{5}{14}\end{pmatrix}\begin{pmatrix} 0 & 0 & 1/2 & 1/2 \\ 0 & 0 & 1/3 & 2/3 \\ 1/4 & 1/4 & 0 & 1/2 \\ 1/5 & 2/5 & 2/5 & 0 \end{pmatrix} = \begin{pmatrix}\frac{2}{14} & \frac{3}{14} & \frac{4}{14} & \frac{5}{14}\end{pmatrix}.$$

In addition to being ergodic, Markov chains associated with networks have another property called **reversibility**. An ergodic chain is said to be **reversible** if $w_x P_{xy} = w_y P_{yx}$ for all x, y. That this is true for our network chains follows from the fact that

$$C_x P_{xy} = C_x \frac{C_{xy}}{C_x} = C_{xy} = C_{yx} = C_y \frac{C_{yx}}{C_y} = C_y P_{yx}.$$

Thus, dividing the first and last term by C, we have $w_x P_{xy} = w_y P_{yx}$.

To see the meaning of reversibility, we start our Markov chain with initial probabilities **w** (in equilibrium) and observe a few states, for example

$$a \quad c \quad b \quad d.$$

The probability that this sequence occurs is

$$w_a P_{ac} P_{cb} P_{bd} = \frac{2}{14} \cdot \frac{1}{2} \cdot \frac{1}{4} \cdot \frac{2}{3} = \frac{1}{84}.$$

The probability that the reversed sequence

$$d \quad b \quad c \quad a$$

occurs is

$$w_d P_{db} P_{bc} P_{ca} = \frac{5}{14} \cdot \frac{2}{5} \cdot \frac{1}{3} \cdot \frac{1}{4} = \frac{1}{84}.$$

Thus the two sequences have the same probability of occurring.

In general, when a reversible Markov chain is started in equilibrium, probabilities for sequences in the correct order of time are the same as those with time reversed. Thus, from data, we would never be able to tell the direction of time.

If **P** is any reversible ergodic chain, then **P** is the transition matrix for a random walk on an electric network; we have only to define $C_{xy} = w_x P_{xy}$. Note, however, if $P_{xx} \neq 0$ the resulting network will need a conductance from x to x (see Exercise 4). Thus reversibility characterizes those ergodic chains that arise from electrical networks. This has to do with the fact that the physical laws that govern the behavior of steady electric currents are invariant under time-reversal (see Onsager [**27**]).

When all the conductances of a network are equal, the associated random walk on the graph G of the network has the property that, from each point, there is an equal probability of moving to each of the points connected to this point by an edge. We shall refer to this random walk as **simple random walk** on G. Most of the examples we have considered so far are simple random walks. Our first example of a random walk on Madison Avenue corresponds to simple random walk on the graph with points $0, 1, 2, \ldots, N$ and edges the streets connecting these points. Our walks on two dimensional graphs were also simple random walks.

EXERCISES

1. Give an example of an ergodic Markov chain that is not regular. (Hint: a chain with two states will do.)

2. Show that, if **P** is the transition matrix for a random walk determined by an electric network, then the fixed vector **w** is given by $w_x = C_x/C$ where $C_x = \Sigma_y C_{xy}$ and $C = \Sigma_x C_x$.

3. Show that, if **P** is a reversible Markov chain and a, b, c are any three states, then the probability, starting at a, of the cycle *abca* is the same as the probability of the reversed cycle *acba*. That is $P_{ab}P_{bc}P_{ca} = P_{ac}P_{cb}P_{ba}$. Show, more generally, that the probability of going around any cycle in the two different directions is the same. (Conversely, if this **cyclic condition** is satisfied, the process is reversible. For a proof, see Kelly [**14**].)

4. Assume that **P** is a reversible Markov chain with $P_{xx} = 0$ for all x. Define an electric network by $C_{xy} = w_x P_{xy}$. Show that the Markov chain associated with this circuit is **P**. Show that we can allow $P_{xx} > 0$ by allowing a conductance from x to x.

5. For the **Ehrenfest urn model**, there are two urns that together contain N balls. Each second, one of the N balls is chosen at random and moved to the other urn. We form a Markov chain with states the number of balls in one of the urns. For $N = 4$, the resulting transition matrix is

$$\mathbf{P} = \begin{array}{c} \\ 0 \\ 1 \\ 2 \\ 3 \\ 4 \end{array} \begin{array}{c} \begin{array}{ccccc} 0 & 1 & 2 & 3 & 4 \end{array} \\ \begin{pmatrix} 0 & 1 & 0 & 0 & 0 \\ 1/4 & 0 & 3/4 & 0 & 0 \\ 0 & 1/2 & 0 & 1/2 & 0 \\ 0 & 0 & 3/4 & 0 & 1/4 \\ 0 & 0 & 0 & 1 & 0 \end{pmatrix} \end{array}.$$

Show that the fixed vector **w** is the binomial distribution **w** = (1/16, 4/16, 6/16, 4/16, 1/16). Determine the electric network associated with this chain.

3.2 Voltages for General Networks; Probabilistic Interpretation

We assume that we have a network of resistors assigned to the edges of a connected graph. We choose two points a and b and put a one-volt battery across these points establishing a voltage $v_a = 1$ and $v_b = 0$, as illustrated in Figure 3.5. We are interested in finding the voltages v_x and the currents i_{xy} in the circuit and in giving a probabilistic interpretation to these quantities.

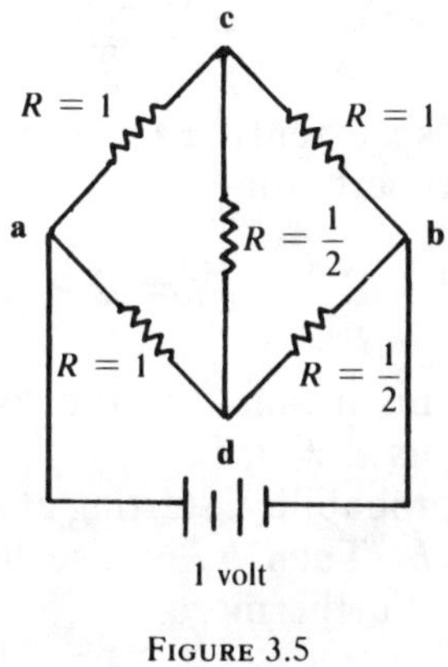

FIGURE 3.5

We begin with the probabilistic interpretation of voltage. It will come as no surprise that we will interpret the voltage as a hitting probability, observing that both functions are harmonic and that they have the same boundary values.

By Ohm's Law, the currents through the resistors are determined by the voltages by

$$i_{xy} = \frac{v_x - v_y}{R_{xy}} = (v_x - v_y)C_{xy}.$$

Note that $i_{xy} = -i_{yx}$. Kirchhoff's Current Law requires that the total current flowing into any point other than a or b is 0. That is, for $x \neq a,b$

$$\sum_y i_{xy} = 0.$$

This will be true if

$$\sum_y (v_x - v_y)C_{xy} = 0$$

or

$$v_x \sum_y C_{xy} = \sum_y C_{xy} v_y.$$

Thus Kirchhoff's Current Law requires that our voltages have the property that

$$v_x = \sum_y \frac{C_{xy}}{C_x} v_y = \sum_y P_{xy} v_y$$

for $x \neq a, b$. This means that the voltage v_x is harmonic at all points $x \neq a, b$.

Let h_x be the probability, starting at x, that state a is reached before b. Then h_x is also harmonic at all points $x \neq a, b$. Furthermore

$$v_a = h_a = 1$$

and

$$v_b = h_b = 0.$$

Thus if we modify **P** by making a and b absorbing states, we obtain an absorbing Markov chain $\bar{\mathbf{P}}$ and v

and h are both solutions to the Dirichlet problem for the Markov chain with the same boundary values. Hence $v = h$.

For our example, the transition probabilities $\bar{P}_{xy}$ are shown in Figure 3.6. The function v_x is harmonic for $\bar{P}$ with boundary values $v_a = 1$, $v_b = 0$.

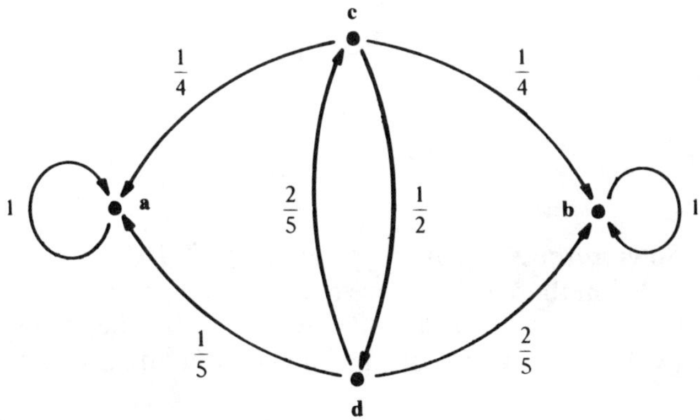

FIGURE 3.6

To sum up, we have the following:

INTREPRETATION OF VOLTAGE. *When a unit voltage is applied between a and b, making $v_a = 1$ and $v_b = 0$, the voltage v_x at any point x represents the probability that a walker starting from x will return to a before reaching b.*

In this probabilistic interpretation of voltage, we have assumed a unit voltage, but we could have assumed an arbitrary voltage v_a between a and b. Then the hitting probability h_x would be replaced by an expected value in a game where the player starts at x and is paid v_a if a is reached before b and 0 otherwise.

Let's use this interpretation of voltage to find the voltages for our example. Referring back to Figure 3.6, we see that

$$v_a = 1$$
$$v_b = 0$$
$$v_c = \frac{1}{4} + \frac{1}{2} v_d$$
$$v_d = \frac{1}{5} + \frac{2}{5} v_c .$$

Solving these equations yields $v_c = 7/16$ and $v_d = 3/8$. From these voltages we obtain the current i_{xy}. For example $i_{cd} = ((7/16) - (3/8)) \cdot 2 = 1/8$. The resulting voltages and currents are shown in Figure 3.7. The

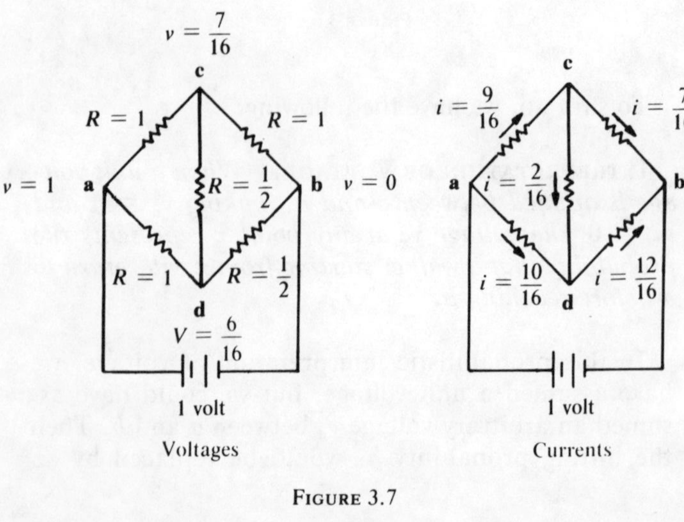

FIGURE 3.7

voltage at c is $7/16$ and so this is also the probability, starting at c, of reaching a before b.

3.3 Probabilistic Interpretation of Current

We turn now to the probabilistic interpretation of current. This interpretation is found by taking a naive view of the process of electrical conduction: We imagine that positively charged particles enter the network at point a and wander around from point to point until they finally arrive at point b, where they leave the network. (It would be more realistic to imagine negatively charged particles entering at b and leaving at a, but realism is not what we're after.) To determine the current i_{xy} along the branch from x to y, we consider that in the course of its peregrinations the point may pass once or several times along the branch from x to y, and in the opposite direction from y to x. We may now hypothesize that the current i_{xy} is proportional to the expected net number of movements along the edge from x to y, where movements from y back to x are counted as negative. This hypothesis is correct, as we will now show.

The walker begins at a and walks until he reaches b; note that if he returns to a before reaching b, he keeps on going. Let u_x be the expected number of visits to state x before reaching b. Then $u_b = 0$ and, for $x \neq a, b$,

$$u_x = \sum_y u_y P_{yx}.$$

This last equation is true because, for $x \neq a, b$, every entrance to x must come from some y.

We have seen that $C_x P_{xy} = C_y P_{yx}$; thus

$$u_x = \sum_y u_y \frac{P_{xy} C_x}{C_y}$$

or

$$\frac{u_x}{C_x} = \sum_y P_{xy} \frac{u_y}{C_y}.$$

This means that $v_x = u_x/C_x$ is harmonic for $x \neq a, b$. We have also $v_b = 0$ and $v_a = u_a/C_a$. This implies that v_x is the voltage at x when we put a battery from a to b that establishes a voltage u_a/C_a at a and a voltage 0 at b. (We remark that the expression $v_a = u_a/C_a$ may be understood physically by viewing u_a as charge and C_a as capacitance; see Kelly [**14**] for more about this.)

We are interested in the current that flows from x to y. This is

$$i_{xy} = (v_x - v_y)C_{xy} = \left(\frac{u_x}{C_x} - \frac{u_y}{C_y}\right)C_{xy}$$

$$= \frac{u_x C_{xy}}{C_x} - \frac{u_y C_{yx}}{C_y} = u_x P_{xy} - u_y P_{yx}.$$

Now $u_x P_{xy}$ is the expected number of times our walker will go from x to y and $u_y P_{yx}$ is the expected number of times he will go from y to x. Thus the current i_{xy} is the expected value for the net number of times the walker passes along the edge from x to y. Note that for any particular walk this net value will be an integer, but the expected value will not.

As we have already noted, the currents i_{xy} here are not those of our original electrical problem, where we apply a 1-volt battery, but they are proportional to those original currents. To determine the constant of proportionality, we note the following characteristic property of the new currents i_{xy}: The total current flowing into the network at a (and out at b) is 1. In symbols,

$$\sum_y i_{ay} = 1.$$

Indeed, from our probabilistic interpretation of i_{xy}, this sum represents the expected value of the difference between the number of times our walker leaves a and enters a. This number is necessarily one and so the current flowing into a is 1.

This unit current flow from a to b can be obtained from the currents in the original circuit, corresponding to a 1-volt battery, by dividing through by the total amount of current

$$\sum_y i_{ay}$$

flowing into a; doing this to the currents in our example yields the unit current flow shown in Figure 3.8.

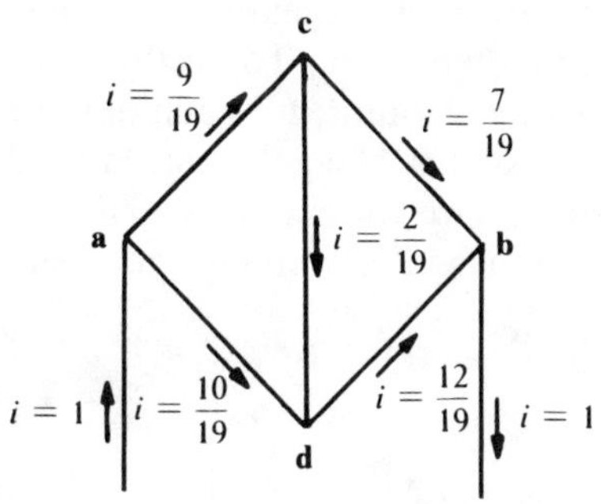

The unit current flow

FIGURE 3.8

This shows that the constant of proportionality we were seeking to determine is the reciprocal of the amount of current that flows through the circuit when a 1-volt battery is applied between a and b. This quantity, called the effective resistance between a and b, is discussed in detail in the next section.

To sum up, we have the following:

INTERPRETATION OF CURRENT. *When a unit current flows into a and out of b, the current i_{xy} flowing through the branch connecting x to y is equal to the expected net number of times that a walker, starting at a and walking until he reaches b, will move along the branch from x to y. These currents are proportional to the currents that arise when a unit voltage is applied between a and b, the constant of proportionality being the effective resistance of the network.*

We have seen that we can estimate the voltages by simulation. We can now do the same for the currents. We have to estimate the expected value for the net number of crossings of xy. To do this, we start a large number of walks at a and, for each one, record the net number of crossings of each edge and then average these as an estimate for the expected value. Carrying out 10,000 such walks yielded the results shown in Figure 3.9.

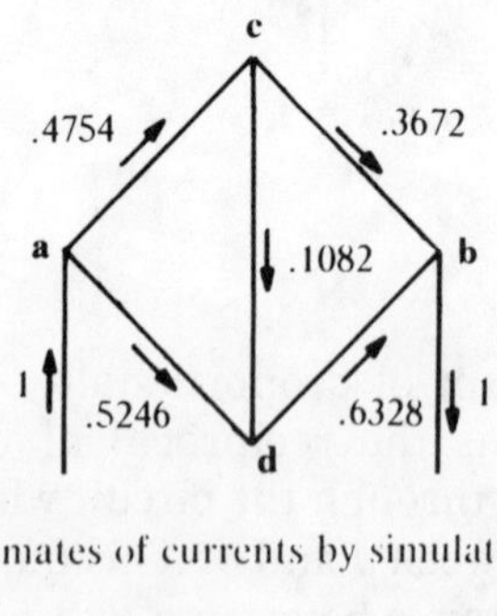

Estimates of currents by simulation

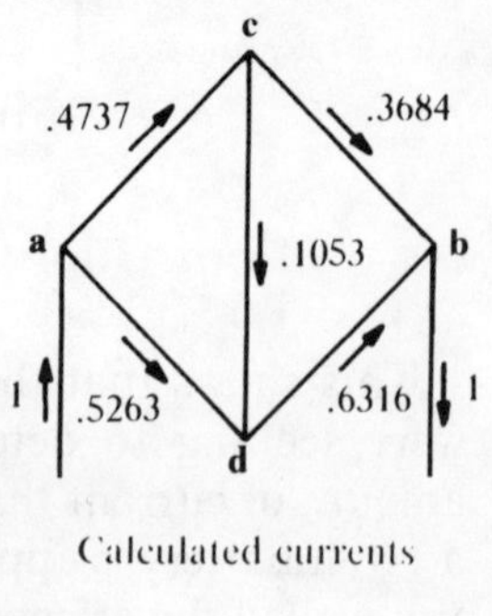

Calculated currents

FIGURE 3.9

The results of simulation are in good agreement with the theoretical values of current. As was the case for estimates of the voltages by simulation, we have statistical errors. Our estimates have the property that the total current flowing into a is 1, out of b is 1, and into any other point it is 0. This is no accident; the explanation is that the history of each walk would have these properties, and these properties are not destroyed by averaging.

EXERCISE

1. Kingman [**18**] introduced a different model for current flow. Kelly [**14**] gave a new interpretation of this model. Both authors use continuous time. A discrete time version of Kelly's interpretation would be the following: At each point of the graph there is a black or a white button. Each second an edge is chosen; edge xy is chosen with probability C_{xy}/C where C is the sum of the conductances. The buttons on the edge chosen are then interchanged. When a button reaches a it is painted black, and when a button reaches b it is painted white. Show that there is a limiting probability p_x that site x has a black button and that p_x is the voltage v_x at x when a unit voltage is imposed between a and b. Show that the current i_{xy} is proportional to the net flow of black buttons along the edge xy. Does this suggest a hypothesis about the behavior of conduction electrons in metals?

3.4 Effective Resistance and the Escape Probability

When we impose a voltage v between points a and b, a voltage $v_a = v$ is established at a and $v_b = 0$, and a current $i_a = \Sigma_x i_{ax}$ will flow into the circuit from the outside source. The amount of current that flows depends upon the overall resistance in the circuit. We define the **effective resistance** R_{EFF} between a and b by

$R_{EFF} = v_a/i_a$. The reciprocal quantity $C_{EFF} = 1/R_{EFF} = i_a/v_a$ is the **effective conductance**. If the voltage between a and b is multiplied by a constant, then the currents are multiplied by the same constant, so R_{EFF} depends only on the ratio of v_a to i_a.

Let us calculate R_{EFF} for our example. When a unit voltage was imposed, we obtained the currents shown in Figure 3.7. The total current flowing into the circuit is $i_a = 9/16 + 10/16 = 19/16$. Thus the effective resistance is

$$R_{EFF} = \frac{v_a}{i_a} = \frac{1}{\frac{19}{16}} = \frac{16}{19}.$$

We can interpret the effective conductance probabilistically as an **escape probability**. When $v_a = 1$, the effective conductance equals the total current i_a flowing into a. This current is

$$\begin{aligned} i_a &= \sum_y (v_a - v_y)C_{ay} = \sum_y (v_a - v_y)\frac{C_{ay}}{C_a}C_a \\ &= C_a(1 - \sum_y P_{ay}v_y) \\ &= C_a p_{\text{escape}} \end{aligned}$$

where p_{escape} is the probability, starting at a, that the walk reaches b before returning to a. Thus

$$C_{EFF} = C_a p_{\text{escape}}$$

and

$$p_{\text{escape}} = \frac{C_{EFF}}{C_a}.$$

In our example $C_a = 2$ and we found that $i_a = 19/16$. Thus

$$p_{\text{escape}} = \frac{19}{32}.$$

In calculating effective resistances, we shall use two important facts about electric networks. First, if two resistors are connected in series, they may be replaced by a single resistor whose resistance is the sum of the

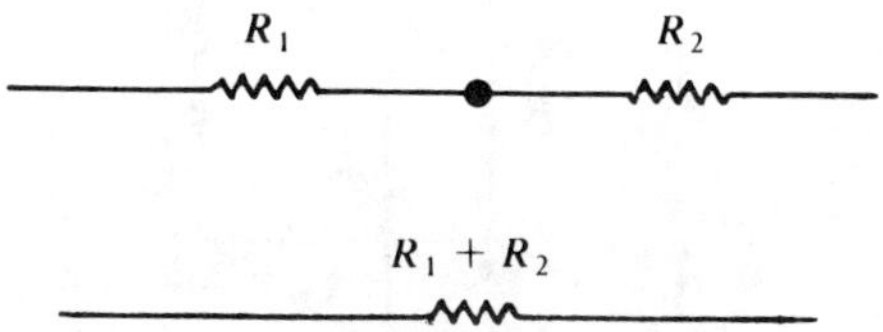

FIGURE 3.10

two resistances. Secondly, two resistors in parallel may be replaced by a single resistor with resistance R such that

$$\frac{1}{R} = \frac{1}{R_1} + \frac{1}{R_2} = \frac{R_1 R_2}{R_1 + R_2}.$$

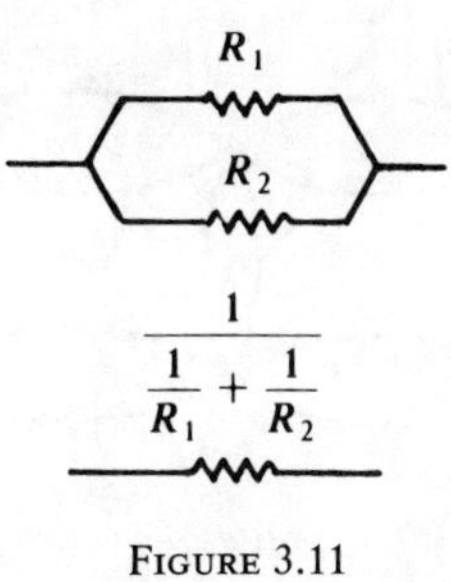

FIGURE 3.11

The second rule can be stated more simply in terms of conductances: If two resistors are connected in parallel, they may be replaced by a single resistor whose conductance is the sum of the two conductances.

We illustrate the use of these ideas to compute the effective resistance between two adjacent points of a unit cube of unit resistors, as shown in Figure 3.12.

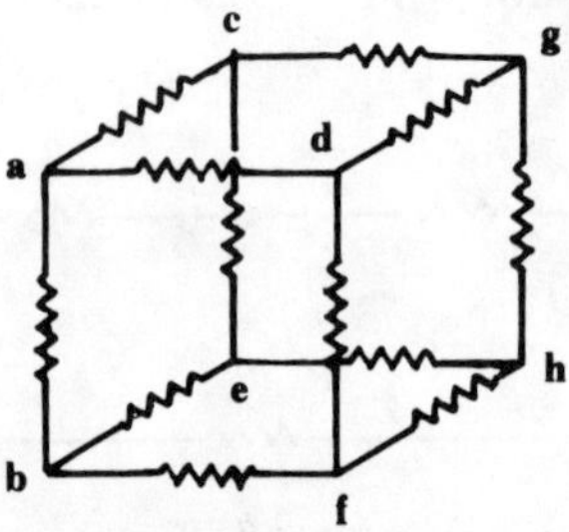

FIGURE 3.12

We put a unit battery between a and b. Then, by symmetry, the voltages at c and d will be the same as will those at e and f. Thus our circuit is equivalent to the circuit shown in Figure 3.13.

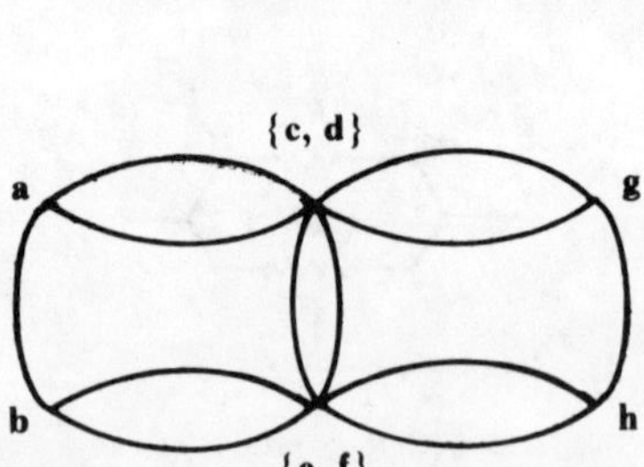

FIGURE 3.13

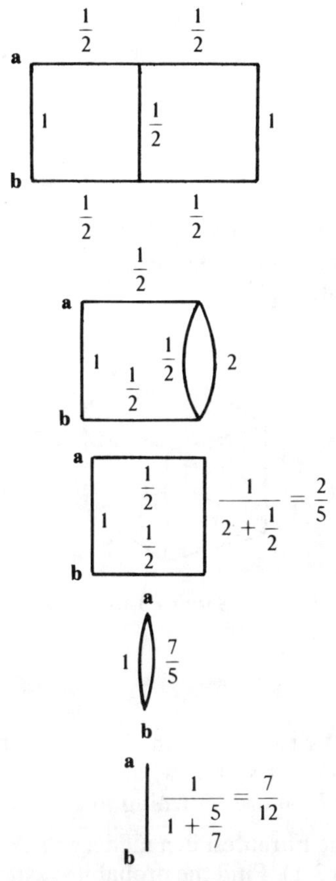

FIGURE 3.14

Using the laws for the effective resistance of resistors in series and parallel, this network can be successively reduced to a single resistor of resistance 7/12 ohms, as shown in Figure 3.14. Thus the effective resistance is 7/12. The current flowing into a from the battery will be $i_a = 1/R_{EFF} = 12/7$. The probability that a walk starting at a will reach b before returning to a is

$$p_{\text{escape}} = \frac{i_a}{C_a} = \frac{\frac{12}{7}}{3} = \frac{4}{7}.$$

This example and many other interesting connections between electric networks and graph theory may be found in Bollobás [**2**].

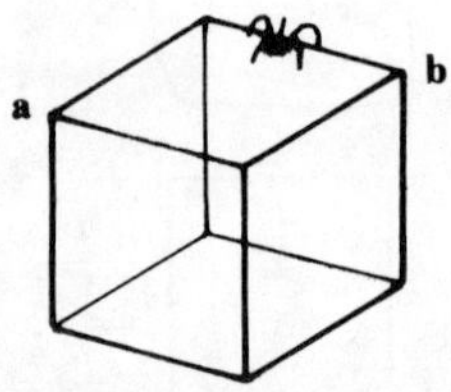

FIGURE 3.15

EXERCISES

1. A bug walks randomly on the unit cube (see Figure 3.15). If the bug starts at a, what is the probability that it reaches food at b before returning to a?

2. Consider the Ehrenfest urn model with $N = 4$ (see Exercise 5 of Section 3.1). Find the probability, starting at 0, that state 4 is reached before returning to 0.

3. Consider the ladder network shown in Figure 3.16. Show that if R_n is the effective resistance of a ladder with n rungs then $R_1 = 2$ and

$$R_{n+1} = \frac{2 + 2R_n}{2 + R_n}.$$

Use this to show that $\lim_{n\to\infty} R_n = \sqrt{2}$.

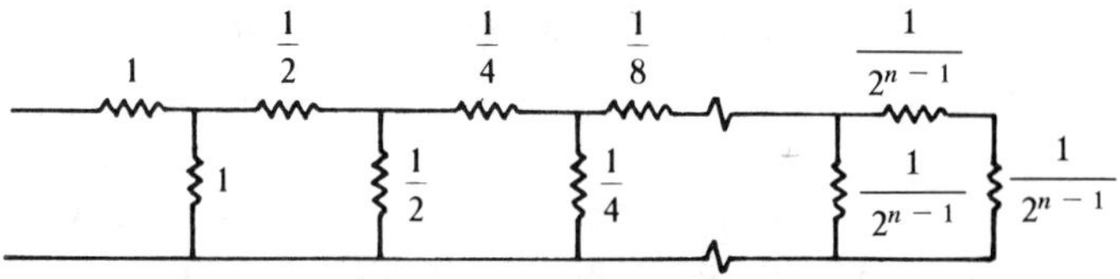

FIGURE 3.16

4. A drunken tourist starts at his hotel and walks at random through the streets of the idealized Paris shown in Figure 3.17. Find the probability that he reaches the Arc de Triomphe before he reaches the outskirts of town.

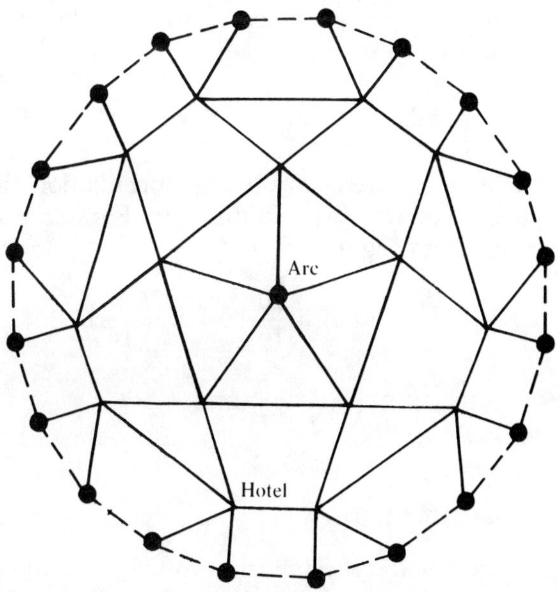

FIGURE 3.17

3.5 Currents Minimize Energy Dissipation

We have seen that when we impose a voltage between a and b voltages v_x are established at the points and currents i_{xy} flow through the resistors. In this section we shall give a characterization of the currents in terms of a quantity called **energy dissipation**. When a current i_{xy} flows through a resistor, the energy dissipated is

$$i_{xy}^2 R_{xy};$$

this is the product of the current i_{xy} and the voltage $v_{xy} = i_{xy}R_{xy}$. The **total energy dissipation** in the circuit is

$$E = \frac{1}{2} \sum_{x,y} i_{xy}^2 R_{xy}.$$

Since $i_{xy}R_{xy} = v_x - v_y$ we can also write the energy dissipation as

$$E = \frac{1}{2} \sum_{x,y} i_{xy}(v_x - v_y).$$

The factor 1/2 is necessary in this formulation since each edge is counted twice in this sum. For our example, we see from Figure 3.7 that

$$\begin{aligned} E &= \left(\frac{9}{16}\right)^2 \cdot 1 + \left(\frac{10}{16}\right)^2 \cdot 1 + \left(\frac{7}{16}\right)^2 \cdot 1 \\ &\quad + \left(\frac{2}{16}\right)^2 \cdot \frac{1}{2} + \left(\frac{12}{16}\right)^2 \cdot \frac{1}{2} \\ &= \frac{19}{16}. \end{aligned}$$

If a source (battery) establishes voltages v_a and v_b at a and b, then the energy supplied is $(v_a - v_b)i_a$ where

$i_a = \Sigma_x i_{ax}$. By conservation of energy, we would expect this to be equal to the energy dissipated. In our example $v_a - v_b = 1$ and $i_a = 19/16$, so this is the case. We shall show that this is true in a somewhat more general setting.

Define a **flow j** from a to b to be an assignment of numbers j_{xy} to pairs xy such that

(a) $j_{xy} = -j_{yx}$

(b) $\sum_y j_{xy} = 0 \qquad$ if $x \neq a, b$

(c) $j_{xy} = 0 \qquad$ if x and y are not adjacent.

We denote by $j_x = \Sigma_y j_{xy}$ the flow into x from the outside. By (b) $j_x = 0$ for $x \neq a, b$. Of course $j_b = -j_a$. To verify this, note that

$$j_a + j_b = \sum_x j_x = \sum_x \sum_y j_{xy}$$

$$= \frac{1}{2} \sum_{x,y} (j_{xy} + j_{yx}) = 0,$$

since $j_{xy} = -j_{yx}$.

With this terminology, we can now formulate the following version of the principle of conservation of energy:

Conservation of Energy. *Let w be any function defined on the points of the graph and* **j** *a flow from a to b. Then*

$$(w_a - w_b)j_a = \frac{1}{2} \sum_{x,y} (w_x - w_y)j_{xy}.$$

Proof.

$$\begin{aligned}\sum_{x,y} (w_x - w_y)j_{xy} &= \sum_x (w_x \sum_y j_{xy}) - \sum_y (w_y \sum_x j_{xy}) \\ &= w_a \sum_y j_{ay} + w_b \sum_y j_{by} \\ &\quad - w_a \sum_x j_{xa} - w_b \sum_x j_{xb} \\ &= w_a j_a + w_b j_b - w_a(-j_a) \\ &\quad - w_b(-j_b) \\ &= 2(w_a - w_b)j_a .\end{aligned}$$

Thus

$$(w_a - w_b)j_a = \frac{1}{2} \sum_{x,y} (w_x - w_y)j_{xy}$$

as was to be proven.

If we now impose a voltage v_a between a and b with $v_b = 0$, we obtain voltages v_x and currents i_{xy}. The currents $\mathbf{i}$ give a flow from a to b and so by the previous result, we conclude that

$$v_a i_a = \frac{1}{2} \sum_{x,y} (v_x - v_y)i_{xy} = \frac{1}{2} \sum_{x,y} i_{xy}^2 R_{xy} .$$

Recall that $R_{EFF} = v_a / i_a$. Thus in terms of resistances we can write this as

$$i_a^2 R_{EFF} = \frac{1}{2} \sum_{x,y} i_{xy}^2 R_{xy} .$$

If we adjust v_a so that $i_a = 1$, we call the resulting flow **the unit current flow** from a to b. The unit current flow from a to b is a particular example of a **unit flow** from a to b, which we define to be any flow i_{xy} from a to b for which $i_a = -i_b = 1$. The formula above shows that the energy dissipated by the unit current flow is just R_{EFF}. According to a basic result called Thomson's Principle, this value is smaller than the energy dissipated by any other unit flow from a to b. Before proving this principle, let us watch it in action in the example worked out above.

Recall that, for this example, we found the true values and some approximate values for the unit current flow; these were shown in Figure 3.9. The energy dissipation for the true currents is

$$E = R_{EFF} = \frac{16}{19} = .8421053.$$

Our approximate currents also form a unit flow and, for these, the energy dissipation is

$$\begin{aligned} \bar{E} &= (.4754)^2 \cdot 1 + (.5246)^2 \cdot 1 + (.3672)^2 \cdot 1 \\ &\quad + (.1082)^2 \cdot \frac{1}{2} + (.6328)^2 \cdot \frac{1}{2} \\ &= .8421177. \end{aligned}$$

We note that $\bar{E}$ is greater than E, though just barely.

Thomson's Principle. (Thomson [35]). *If* $\mathbf{i}$ *is the unit flow from a to b determined by Kirchhoff's Law, then the energy dissipation* $1/2\ \Sigma_{x,y}\, i_{xy}^2 R_{xy}$ *minimizes the energy dissipation* $1/2\ \Sigma_{x,y}\, j_{xy}^2 R_{xy}$ *among all unit flows* $\mathbf{j}$ *from a to b.*

Proof. Let $\mathbf{j}$ be any unit flow from a to b and let $d_{xy} = j_{xy} - i_{xy}$. Then $\mathbf{d}$ is a flow from a to b with $d_a = \Sigma_x d_{ax} = 1 - 1 = 0$.

$$\sum_{x,y} j_{xy}^2 R_{xy} = \sum_{x,y} (i_{xy} + d_{xy})^2 R_{xy}$$
$$= \sum_{x,y} i_{xy}^2 R_{xy} + 2\sum_{x,y} i_{xy} R_{xy} d_{xy} + \sum_{x,y} d_{xy}^2 R_{xy}$$
$$= \sum_{x,y} i_{xy}^2 R_{xy} + 2\sum_{x,y} (v_x - v_y) d_{xy} + \sum_{x,y} d_{xy}^2 R_{xy}.$$

Setting $\mathbf{w} = \mathbf{v}$ and $\mathbf{j} = \mathbf{d}$ in the conservation of energy result above shows that the middle term is $4(v_a - v_b)d_a = 0$. Thus

$$\sum_{x,y} j_{xy}^2 R_{xy} = \sum_{x,y} i_{xy}^2 R_{xy} + \sum_{x,y} d_{xy}^2 R_{xy} \geqslant \sum_{x,y} i_{xy}^2 R_{xy}.$$

This completes the proof.

Exercises

1. The following is the so-called "dual form" of Thomson's Principle. Let u be any function on the points of the graph G of a circuit such that $u_a = 1$ and $u_b = 0$. Then the energy dissipation

$$\frac{1}{2} \sum_{x,y} (u_x - u_y)^2 C_{xy}$$

is minimized by the voltages v_x that result when a unit voltage is established between a and b, i.e., $v_a = 1$, $v_b = 0$, and the other voltages are determined by Kirchhoff's Laws. Prove this dual principle. This second principle is known nowadays as the **Dirichlet Principle**, though it too was discovered by Thomson.

2. In Section 2.4 we stated that, to solve the Dirichlet problem by the method of relaxations, we could start with an arbitrary initial guess. Show that when we replace the value at a point by the average of the neighboring points the energy dissipation, as expressed in Exercise 1, can only decrease. Use this to prove that the relaxation method converges to a solution of the Dirichlet problem for an arbitrary initial guess.

CHAPTER 4

RAYLEIGH'S MONOTONICITY LAW

4.1 Rayleigh's Monotonicity Law

In this chapter we will study Rayleigh's Monotonicity Law. This law from electric network theory will be an important tool in our future study of random walks. In this first section we will give an example of the use of this law.

Consider a random walk on streets of a city as in Figure 4.1. Let p_{escape} be the probability that a walker starting from a reaches b before returning to a. Assign to each edge a unit resistance and maintain a voltage of

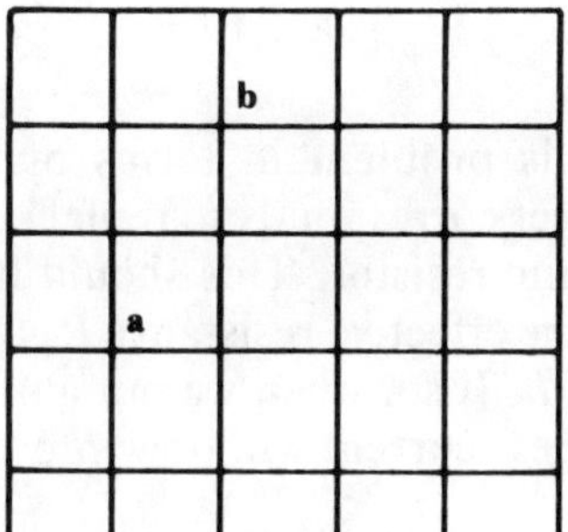

FIGURE 4.1

one volt between a and b; then a current i_a will flow into the circuit and we showed in Section 3.4 that

$$p_{\text{escape}} = \frac{i_a}{C_a} = \frac{i_a}{4}.$$

Now suppose that one of the streets (not connected to a) becomes blocked. Our walker must choose from the remaining streets if he reaches a corner of this street. The modified graph will be as in Figure 4.2. We want to show that the probability of escaping to b from a is decreased.

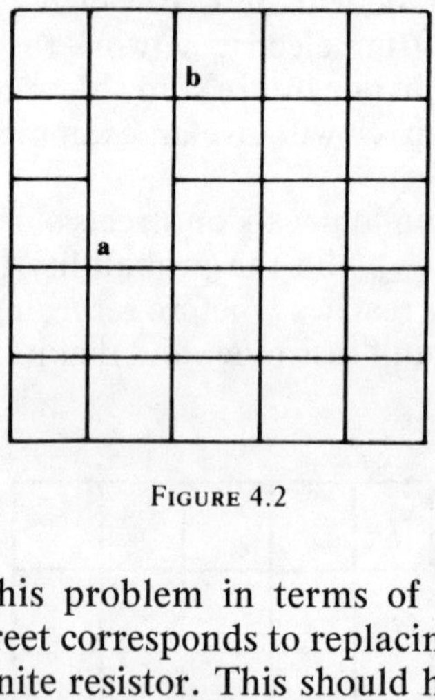

FIGURE 4.2

Consider this problem in terms of our network. Blocking a street corresponds to replacing a unit resistor by an infinite resistor. This should have the effect of increasing the effective resistance R_{EFF} of the circuit between a and b. If so, when we put a unit voltage between a and b less current will flow into the circuit and

$$p_{\text{escape}} = \frac{i_a}{4} = \frac{1}{4\,R_{EFF}}.$$

will decrease.

Thus we need only show that when we increase the resistance in one part of a circuit, the effective resistance increases. This fact, known as Rayleigh's Monotonicity Law, is almost self-evident. Indeed, the father of electromagnetic theory, James Clerk Maxwell, regarded this to be the case. In his *Treatise on Electricity and Magnetism* ([**23**], p. 427), he wrote

"If the specific resistance of any portion of the conductor be changed, that of the remainder being unchanged, the resistance of the whole conductor will be increased if that of the portion is increased, and diminished if that of the portion is diminished. This principle may be regarded as self-evident. . . ."

Rayleigh's Monotonicity Law. *If the resistances of a circuit are increased, the effective resistance* R_{EFF} *between any two points can only increase. If they are decreased, it can only decrease.*

Proof. Let **i** be the unit current flow from a to b with the resistors R_{xy}. Let **j** be the unit current flow from a to b with the resistors $\bar{R}_{xy}$ with $\bar{R}_{xy} \geqslant R_{xy}$. Then

$$\begin{aligned} \bar{R}_{EFF} &= \frac{1}{2} \sum_{x,y} j_{xy}^2 \bar{R}_{xy} \\ &\geqslant \frac{1}{2} \sum_{x,y} j_{xy}^2 R_{xy}. \end{aligned}$$

But since **j** is a unit flow from a to b, Thomson's Principle tells us that the energy dissipation, calculated with resistors R_{xy}, is bigger than that for the true currents determined by these resistors: that is

$$\frac{1}{2} \sum_{x,y} j_{xy}^2 R_{xy} \geqslant \frac{1}{2} \sum_{x,y} i_{xy}^2 R_{xy} = R_{EFF}$$

Thus $\bar{R}_{EFF} \geqslant R_{EFF}$. The proof for the case of decreasing resistances is the same.

Exercises

1. Consider a graph G and let R_{xy} and $\bar{R}_{xy}$ be two different assignments of resistances to the edges of G. Let $\bar{\bar{R}}_{xy} = \bar{R}_{xy} + R_{xy}$. Let R_{EFF}, $\bar{R}_{EFF}$ and $\bar{\bar{R}}_{EFF}$ be the effective resistances when R, $\bar{R}$ and $\bar{\bar{R}}$, respectively, are used. Prove that

$$\bar{\bar{R}}_{EFF} \geqslant \bar{R}_{EFF} + R_{EFF}.$$

Conclude that the effective resistance of a network is a concave function of any of its component resistances (Shannon and Hagelbarger [33]).

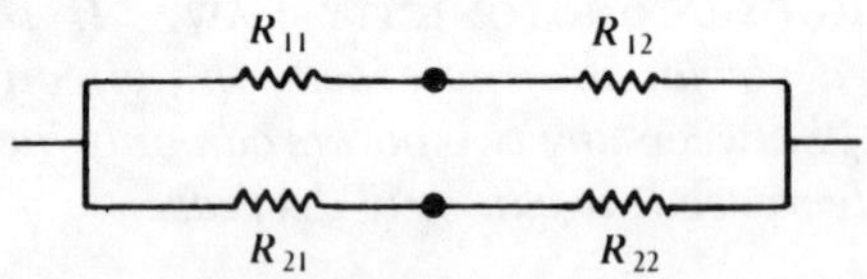

FIGURE 4.3

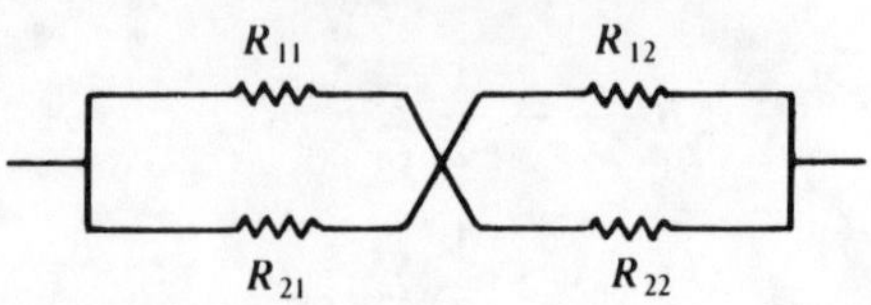

FIGURE 4.4

2. Show that the effective resistance of the circuit in Figure 4.3 is greater than or equal to the effective resistance of the circuit in Figure 4.4. Use this to show the following inequality for $R_{ij} \geqslant 0$:

$$\frac{1}{\dfrac{1}{R_{11}+R_{12}}+\dfrac{1}{R_{21}+R_{22}}} \geqslant \frac{1}{\dfrac{1}{R_{11}}+\dfrac{1}{R_{21}}}+\frac{1}{\dfrac{1}{R_{12}}+\dfrac{1}{R_{22}}}.$$

See the note by Lehman [**20**] for a proof of the general Minkowski inequality by this method.

3. Let **P** be the transition matrix associated with an electric network and let a, b, r, s be four points on the network. Let $\bar{\mathbf{P}}$ be a transition matrix defined on the state space $S = \{a, b, r, s\}$. Let $\bar{P}_{ii} = 0$ and for $i \neq j$ define $\bar{P}_{ij}$ to be the probability that, if the chain P is started in state i, then the next time it is in the set $S - \{i\}$ it is in the state j. Show that $\bar{\mathbf{P}}$ is a reversible Markov chain and hence corresponds to an electric network of the form of a Wheatstone Bridge, shown in Figure 4.5. Explain how this proves that, in order to prove the Monotonicity Law, it is sufficient to prove that R_{EFF} is a monotone function of the component resistances for a Wheatstone Bridge. Give a direct proof of the Monotonicity Law for this special case.

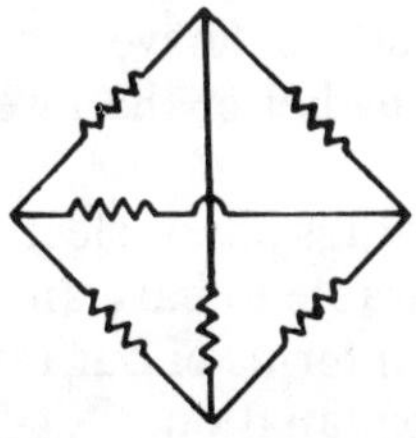

FIGURE 4.5

4.2 A Probabilistic Explanation of the Monotonicity Law

We have quoted Maxwell's assertion that Rayleigh's Monotonicity Law may be regarded as self-evident, but one might feel that any argument in terms of electricity is only self-evident if we know what electricity is. In Cambridge, they tell the following story about Maxwell: Maxwell was lecturing and, seeing a student dozing off, awakened him, asking, "Young man, what is electricity?" "I'm terribly sorry, sir," the student replied, "I knew the answer but I have forgotten it." Maxwell's response to the class was, "Gentlemen, you have just witnessed the greatest tragedy in the history of science. The one person who knew what electricity is has forgotten it."

To say that our intuition about the Monotonicity Law is only as solid as our understanding of electricity is not really a valid argument, of course, because in saying that this law is self-evident we are secretly depending on the analogy between electricity and the flow of water (see Feynman [**6**], Vol. 2, Chapter 12). We just can't believe that if a water main gets clogged the total rate of flow out of the local reservoir is going to increase. But as soon as we admit this, some pedant will ask if we're talking about flows with low Reynolds' number, or what, and we'll have to admit that we don't understand water any better than we understand electricity.

Whatever our feelings about electricity or the flow of water, it seems desirable to have an explanation of the Monotonicity Law in terms of our random walker. We now give such an explanation.

We begin by collecting some notation and results from previous sections. As usual, we have a network of conductances (streets) and a walker who moves from point x to point y with probability

$$P_{xy} = \frac{C_{xy}}{C_x}$$

where C_{xy} is the conductance from x to y and $C_x = \Sigma_y C_{xy}$. We choose two preferred points a and b. The walker starts at a and walks until he reaches b or returns to a. We denote by v_x the probability that the walker, starting at x, reaches a before b. Then $v_a = 1$, $v_b = 0$, and the function v_x is harmonic at all points $x \neq a, b$. We denote by p_{escape} the probability that the walker, starting at a, reaches b before returning to a. Then

$$p_{\text{escape}} = 1 - \sum_x p_{ax} v_x .$$

Now we have seen that the effective conductance between a and b is

$$c_a p_{\text{escape}} .$$

We wish to show that this increases whenever one of the conductances C_{rs} is increased. If a is different from r or s, we need only show that p_{escape} increases. The case where r or s coincides with a is easily disposed of (see Exercise 1). The case where r or s coincides with b is also easy (see Exercise 2). Hence from now on we will assume that $r, s \neq a$ and $r, s \neq b$.

Instead of increasing C_{rs}, we can think of adding a new edge of conductance ϵ between r and s. (See Figure 4.6.) We will call this new edge a "bridge" to distinguish it from the other edges. Note that the graph with the bridge added will have more than one edge between r and s (unless there was no edge between r and s in the original graph), and this will complicate any expression that involves summing over edges. Everything we have said or will say in this book holds for graphs with "multiple edges" as well as for graphs without them. So far, we have chosen to keep our expressions simple by assuming that an edge is determined by its

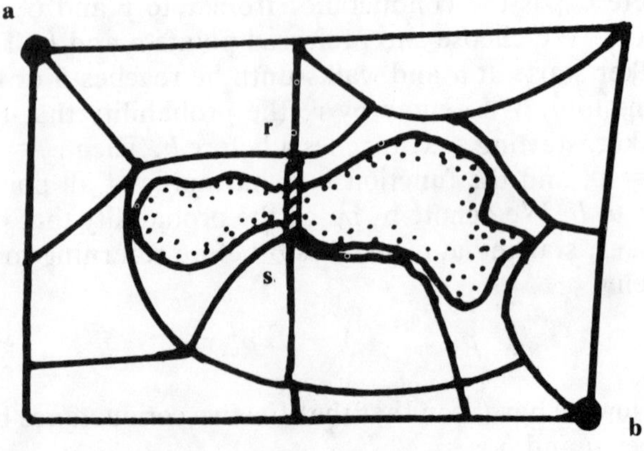

FIGURE 4.6

endpoints. The trade-off is that in the manipulations below, whenever we write a sum over edges we will have to add an extra term to account for the bridge.

Why should adding the bridge increase the escape probability? The first thing you think is, "Of course, it opens up new possibilities of escaping!" The next instant you think, "Wait a minute, it also opens up new possibilities of returning to the starting point. What ensures that the first effect will outweigh the second?" As we shall see, the proper reply is, "Because the walker will cross the bridge more often in the good direction than in the bad direction." To turn this blithe reply into a real explanation will require a little work, however.

To begin to make sense of the notion that the bridge gets used more often in the good direction than the bad, we will make a preliminary argument that applies to any edge of any graph. Let G be any graph, and let rs be any edge with endpoints not a or b. Assume that

$v_r \geqslant v_s$. Since he has a better chance to escape from s than from r, this means that to cross this edge in the good direction is to go from r to s. We shall show that the walker will cross the edge from r to s more often on the average than from s to r.

Let u_x be the expected number of times the walker is at x and u_{xy} the expected number of times he crosses the edge xy from x to y before he reaches b or returns to a. The calculation carried out in Section 3.3 shows that u_x/C_x is harmonic for $x \neq a, b$ with $u_a/C_a = 1/C_a$ and $u_b/C_b = 0$. But the function v_x/C_a also has these properties, so by the Uniqueness Principle

$$\frac{u_x}{C_x} = \frac{v_x}{C_a}.$$

Now

$$u_{rs} = u_r P_{rs} = u_r \frac{C_{rs}}{C_r} = v_r \frac{C_{rs}}{C_a}$$

and

$$u_{sr} = u_s P_{sr} = u_s \frac{C_{sr}}{C_s} = v_s \frac{C_{sr}}{C_a}.$$

Since $C_{rs} = C_{sr}$, and since by assumption $v_r \geqslant v_s$, this means that $u_{rs} \geqslant u_{sr}$. Therefore, we see that any edge leads the walker more often to the more favorable of the points of the edge.

Now let's go back and think about the graph with the bridge. The above discussion shows that the bridge helps in the sense that, on the average, the bridge is crossed more often in the direction that improves the chance of escaping. While this fact is suggestive, it doesn't tell us that we are more likely to escape than if the bridge weren't there; it only tells us what goes on once the bridge is in place. What we need is to make a "before and after" comparison.

Recall that we are denoting the conductance of the bridge by ϵ. To distinguish the quantities pertaining to the walks with and without the bridge, we will put ϵ superscripts on quantities that refer to the walk with the bridge, so that, e.g., $p^{\epsilon}_{\text{escape}}$ denotes the escape probability with the bridge.

Now let d^{ϵ} denote the expected net number of times the walker crosses from r to s. As above, we have

$$\begin{aligned} d^{\epsilon} &= u_r^{\epsilon} \frac{\epsilon}{C_r + \epsilon} - u_s^{\epsilon} \frac{\epsilon}{C_s + \epsilon} \\ &= \left(\frac{u_r^{\epsilon}}{C_r + \epsilon} - \frac{u_s^{\epsilon}}{C_s + \epsilon} \right) \epsilon. \end{aligned}$$

CLAIM: $p^{\epsilon}_{\text{escape}} = p_{\text{escape}} + (v_r - v_s) d^{\epsilon}$.

Why? Every time you use the bridge to go from r to s, you improve your chances of escaping by

$$(1 - v_s) - (1 - v_r) = v_r - v_s,$$

assuming that you would continue your walk without using the bridge. To get the probability of escaping with the bridge, you take the probability of escaping without the bridge, and correct it by adding in the change attributable to the bridge, which is the difference in the original escape probabilities at the ends of the bridge, multiplied by the net number of times you expect to cross the bridge.

Proof. Suppose you're playing a game where you walk around the graph with the bridge, and your fortune when you're at x is v_x, which is the probability that you would return to a before reaching b if the bridge weren't there. You start at a, and walk until you reach b or return to a.

This is not a fair game. Your initial fortune is 1 since you start at a and $v_a = 1$. Your expected final fortune is

$$\begin{aligned} 1 \cdot (1 - p^{\epsilon}_{\text{escape}}) &+ 0 \cdot p^{\epsilon}_{\text{escape}} \\ &= 1 - p^{\epsilon}_{\text{escape}}. \end{aligned}$$

The amount you expect to lose by participating in the game is

$$p^{\epsilon}_{\text{escape}}.$$

(Note that escaping has suddenly become a bad thing!)

Let's analyze where it is that you expect to lose money. First of all, you lose money when you take the first step away from a. The amount you expect to lose is

$$\begin{aligned} 1 - \sum_x P^{\epsilon}_{ax} v_x & \\ = p_{\text{escape}}. \end{aligned}$$

Now if your fortune were given by v^{ϵ}_x instead of v_x, the game would be fair after this first step. However, the function v_x is not harmonic for the walk with the bridge; it fails to be harmonic at r and s. Every time you step away from r, you expect to lose an amount

$$\begin{aligned} v_r - \left(\sum_x \frac{C_{rx}}{C_r + \epsilon} v_x + \frac{\epsilon}{C_r + \epsilon} v_s \right) & \\ = (v_r - v_s) \frac{\epsilon}{C_r + \epsilon}. \end{aligned}$$

Similarly, every time you step away from s you expect to lose an amount

$$(v_s - v_r)\frac{\epsilon}{C_s + \epsilon}.$$

The total amount you expect to lose by participating in the game is:

expected loss at first step
+ (expected loss at r) · (expected number of times at r)
+ (expected loss at s) · (expected number of times at s)

$$= p_{\text{escape}} + (v_r - v_s)\frac{\epsilon}{C_r + \epsilon}u_r^\epsilon + (v_s - v_r)\frac{\epsilon}{C_s + \epsilon}u_s^\epsilon$$

$$= p_{\text{escape}} + (v_r - v_s)d^\epsilon.$$

Equating this with our first expression for the expected cost of playing the game yields the formula we were trying to prove.

According to the formula just established,

$$p^\epsilon_{\text{escape}} - p_{\text{escape}} = (v_r - v_s)\left(\frac{u_r^\epsilon}{C_r + \epsilon} - \frac{u_s^\epsilon}{C_s + \epsilon}\right)\epsilon.$$

For small ϵ, we will have

$$\frac{u_r^\epsilon}{C_r + \epsilon} - \frac{u_s^\epsilon}{C_s + \epsilon} \approx \frac{u_r}{C_r} - \frac{u_s}{C_s}$$

$$= \frac{v_r}{C_a} - \frac{v_s}{C_a},$$

so for small ϵ

$$p^{\epsilon}_{\text{escape}} - p_{\text{escape}} \approx (v_r - v_s)^2 \frac{\epsilon}{C_a}.$$

This approximation allows us to conclude that

$$p^{\epsilon}_{\text{escape}} \geqslant p_{\text{escape}} \geqslant 0$$

for small ϵ. But this is enough to establish the monotonicity law, since any finite change in ϵ can be realized by making an infinite chain of graphs each of which is obtained from the one before by adding a bridge of infinitesimal conductance.

To recapitulate, the difference in the escape probabilities with and without the bridge is obtained by taking the difference between the original escape probabilities at the ends of the bridge, and multiplying by the expected net number of crossings of the bridge. This quantity is positive because the walker tends to cross the bridge more often in the good direction than in the bad direction.

Exercises

1. Give a probabilistic argument to show that $C_a p_{\text{escape}}$ increases with C_{ar} for any r. Give an example to show that p_{escape} by itself may actually decrease.

2. Give a probabilistic argument to show that $C_a p_{\text{escape}}$ increases with C_{rb} for any r.

3. Show that when $v_r = v_s$, changing the value of C_{rs} does not change p_{escape}.

4. Show that

$$\frac{\partial}{\partial R_{rs}} R_{EFF} = i^2_{rs}.$$

5. In this exercise we ask you to derive an exact formula for the change in escape probability

$$p^{\epsilon}_{\text{escape}} - p_{\text{escape}},$$

in terms of quantities that refer only to the walk without the bridge.

(a) Let N_{xy} denote the expected number of times in state y for a walker who starts at x and walks around the graph without the bridge until he reaches a or b. It is a fact that

$$u^{\epsilon}_r = u_r + u^{\epsilon}_r \frac{\epsilon}{C_r + \epsilon}(N_{sr} + 1 - N_{rr})$$

$$+ u^{\epsilon}_s \frac{\epsilon}{C_s + \epsilon}(N_{rr} - N_{sr}).$$

Explain in words why this formula is true.

(b) This equation for u^{ϵ}_r can be rewritten as follows:

$$\frac{C_r}{C_r + \epsilon} u^{\epsilon}_r = u_r + d^{\epsilon}(N_{sr} - N_{rr}).$$

Prove this formula. (Hint: Consider a game where your fortune at x is N_{xr}, and where you start from a and walk on the graph with the bridge until you reach b or return to a.)

(c) Write down the corresponding formula for u^{ϵ}_s, and use this formula to get an expression for d^{ϵ} in terms of quantities that refer to the walk without the bridge.

(d) Use the expression for d^{ϵ} to express $p^{\epsilon}_{\text{escape}} - p_{\text{escape}}$ in terms of quantities that refer to the walk without the bridge, and verify that the value of your expression is $\geqslant 0$ for $\epsilon \geqslant 0$.

6. Give a probabilistic interpretation of the energy dissipation rate.

4.3 A Markov Chain Proof of the Monotonicity Law

Let $\mathbf{P}$ be the ergodic Markov chain associated with an electric network. When we add an ϵ bridge from r to s, we obtain a new transition matrix $\mathbf{P}^{\epsilon}$ that differs from $\mathbf{P}$ only for transitions from r and s. We can minimize the differences between $\mathbf{P}$ and $\mathbf{P}^{\epsilon}$ by replacing $\mathbf{P}$ by the matrix $\hat{\mathbf{P}}$ corresponding to the circuit without the bridge but with an ϵ conductance added from r to r and from s to s. This allows the chain to stay in states r and s but does not change the escape probability from a to b. Thus, we can compare the escape probabilities for the two matrices $\hat{\mathbf{P}}$ and $\mathbf{P}^{\epsilon}$, which differ only by

$$\hat{P}_{rr} = \frac{\epsilon}{C_r + \epsilon} \qquad P^{\epsilon}_{rr} = 0$$

$$\hat{P}_{rs} = \frac{C_{rs}}{C_r + \epsilon} \qquad P^{\epsilon}_{rs} = \frac{C_{rs} + \epsilon}{C_r + s}$$

$$\hat{P}_{ss} = \frac{\epsilon}{C_s + \epsilon} \qquad P^{\epsilon}_{ss} = 0$$

$$\hat{P}_{sr} = \frac{C_{sr}}{C_s + \epsilon} \qquad P^{\epsilon}_{sr} = \frac{C_{sr} + \epsilon}{C_s + \epsilon}.$$

We make states a and b into absorbing states. Let $\hat{\mathbf{N}}$ and $\mathbf{N}^{\epsilon}$ be the fundamental matrices for the absorbing chains obtained from $\hat{\mathbf{P}}$ and $\mathbf{P}^{\epsilon}$ respectively. Then $\hat{\mathbf{N}} = (\mathbf{I} - \hat{\mathbf{Q}})^{-1}$ and $\mathbf{N}^{\epsilon} = (\mathbf{I} - \mathbf{Q}^{\epsilon})^{-1}$ where $\hat{\mathbf{Q}}$ and $\mathbf{Q}^{\epsilon}$ differ only for the four components involving only r and s. That is,

$$\mathbf{Q}^{\epsilon} = \hat{\mathbf{Q}} + \begin{array}{c} \\ r \\ s \end{array}\!\!\begin{array}{c} \begin{array}{cc} \quad r & \qquad\qquad s \end{array} \\ \begin{pmatrix} -\dfrac{\epsilon}{C_r + \epsilon} & \dfrac{\epsilon}{C_r + \epsilon} \\[2ex] \dfrac{\epsilon}{C_s + \epsilon} & -\dfrac{\epsilon}{C_s + \epsilon} \end{pmatrix} \end{array}$$

$$= \hat{\mathbf{Q}} + \mathbf{h}\beta$$

where **h** is the column vector with only components r and s non-zero

$$\mathbf{h} = \begin{matrix} r \\ \\ s \end{matrix}\begin{pmatrix} \dfrac{\epsilon}{C_r + \epsilon} \\ \\ -\dfrac{\epsilon}{C_s + \epsilon} \end{pmatrix}$$

and β is a row vector with only components r and s non-zero

$$\beta = (\quad \overset{r}{-1} \quad \overset{s}{1} \quad).$$

J. G. Kemeny has pointed out to us that if **A** is any matrix with inverse **N** and we add to **A** a matrix of the form $-\mathbf{h}\beta$, then $\mathbf{A} - \mathbf{h}\beta$ has an inverse if and only if $\beta\mathbf{N}\mathbf{h} \neq 1$ and, if so, $\bar{\mathbf{N}} = (\mathbf{A} - \mathbf{h}\beta)^{-1}$ is given by

$$\bar{\mathbf{N}} = \mathbf{N} + c(\mathbf{N}\mathbf{h})(\beta\mathbf{N})$$

where $c = 1/(1 - \beta\mathbf{N}\mathbf{h})$. You are asked to prove this in Exercise 1. Adding $-\mathbf{h}\beta$ to $\mathbf{A} = \mathbf{I} - \hat{\mathbf{Q}}$ and using this result, we obtain

$$\mathbf{N}^{\epsilon} = \hat{\mathbf{N}} + c(\hat{\mathbf{N}}\mathbf{h})(\beta\hat{\mathbf{N}}).$$

Using the simple nature of **h** and β we obtain

$$N_{ij}^{\epsilon} = \hat{N}_{ij} + \left(\frac{\hat{N}_{ir}\epsilon}{C_r + \epsilon} - \frac{\hat{N}_{is}\epsilon}{C_s + \epsilon}\right)(N_{sj} - N_{rj})$$

and

$$c = \frac{1}{1 + \dfrac{\hat{N}_{rr}\epsilon}{C_r + \epsilon} - \dfrac{\hat{N}_{sr}\epsilon}{C_r + \epsilon} + \dfrac{\hat{N}_{ss}\epsilon}{C_s + \epsilon} - \dfrac{\hat{N}_{rs}\epsilon}{C_s + \epsilon}}.$$

Since $\hat{N}_{rr}$ is the expected number of times in r starting in r and $\hat{N}_{sr}$ is the expected number of times in r started in s, $\hat{N}_{rr} \geqslant \hat{N}_{sr}$. Similarly, $\hat{N}_{ss} \geqslant \hat{N}_{rs}$ and so the denominator is $\geqslant 1$. In particular, it is positive.

Recall that the absorption probabilities for state b are given by

$$B_{ib} = \sum_k N_{ik} P_{kb}.$$

Since $P^{\epsilon}_{kb} = \hat{P}_{kb}$,

$$B^{\epsilon}_{ib} = \hat{B}_{ib} + c\left(\frac{\hat{N}_{ir}\epsilon}{C_r + \epsilon} - \frac{\hat{N}_{is}\epsilon}{C_s + \epsilon}\right)(\hat{B}_{sb} - \hat{B}_{rb}).$$

Since $P^{\epsilon}_{aj} = \hat{P}_{aj}$,

$$p^{\epsilon}_{\text{escape}} = \hat{p}_{\text{escape}} + c\left(\frac{\hat{u}_r\epsilon}{C_r + \epsilon} - \frac{\hat{u}_s\epsilon}{C_s + \epsilon}\right)(\hat{B}_{sb} - \hat{B}_{rb}),$$

where $\hat{u}_x$ is the expected number of times that the ergodic chain $\hat{\mathbf{P}}$, started at state a, is in state x before returning to a or reaching b for the first time. The absorption probability B_{xa} is the quantity v_x introduced in the previous section. As shown there, reversibility allows us to conclude that

$$\frac{\hat{u}_x}{\hat{C}_x} = \frac{\hat{B}_{xa}}{\hat{C}_a} = \frac{\hat{B}_{xa}}{C_a}$$

so that

$$p^{\epsilon}_{\text{escape}} = p_{\text{escape}} + \frac{\epsilon c}{C_a}(\hat{B}_{sb} - \hat{B}_{rb})^2$$

and this shows that the Monotonicity Law is true.

The change from $\mathbf{P}$ to $\hat{\mathbf{P}}$ was merely to make the calculations easier. As we have remarked, the escape probabilities are the same for the two chains as are the absorption probabilities B_{ib}. Thus we can remove the hats and write the same formula.

$$p^{\epsilon}_{\text{escape}} = p_{\text{escape}} + \frac{\epsilon c}{C_a}(B_{sb} - B_{rb})^2$$

The only quantity in this final expression that seems to depend upon quantities from $\hat{\mathbf{P}}$ is c. In Exercise 2 you are asked to show that c can also be expressed in terms of the fundamental matrix $\mathbf{N}$ obtained from the original $\mathbf{P}$.

Exercises

1. Let $\mathbf{A}$ be a matrix with inverse $\mathbf{N} = \mathbf{A}^{-1}$. Let $\mathbf{h}$ be a column vector and β a row vector. Show that

$$\bar{\mathbf{N}} = (\mathbf{A} - \mathbf{h}\beta)^{-1}$$

exists if and only if $\beta\mathbf{Nh} \neq 1$ and, if so,

$$\bar{\mathbf{N}} = \mathbf{N} + \frac{(\mathbf{Nh})(\beta\mathbf{N})}{1 - \beta\mathbf{Nh}}.$$

2. Show that c can be expressed in terms of the fundamental matrix $\mathbf{N}$ of the original Markov chain $\mathbf{P}$ by

$$c = \frac{1}{1 + \frac{N_{rr}\epsilon}{C_r} - \frac{N_{sr}\epsilon}{C_r} + \frac{N_{ss}\epsilon}{C_s} - \frac{N_{rs}\epsilon}{C_s}}.$$

PART II

Random Walks on Infinite Networks

CHAPTER 5

PÓLYA'S RECURRENCE PROBLEM

5.1 Random Walks on Lattices

In 1921 George Pólya [**28**] investigated random walks on certain infinite graphs, or as he called them, "street networks." The graphs he considered, which we will refer to as **lattices**, are illustrated in Figure 5.1.

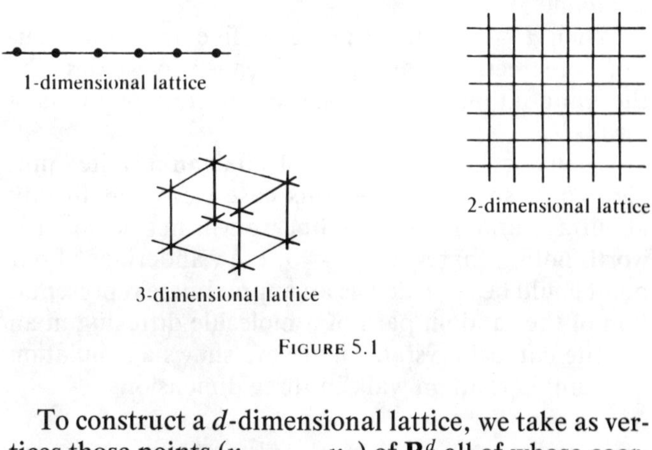

FIGURE 5.1

To construct a d-dimensional lattice, we take as vertices those points $(x_1, \ldots, x_d)$ of $\mathbf{R}^d$ all of whose coordinates are integers, and we join each vertex by an undirected line segment to each of its $2d$ nearest neighbors. These connecting segments, which represent the edges of our graph, each have unit length and run parallel to one of the coordinate axes of $\mathbf{R}^d$. We will denote this d-dimensional lattice by $\mathbf{Z}^d$. We will denote the origin $(0, 0, \ldots, 0)$ by $\mathbf{0}$.

Now let a point walk around at random on this lattice. As usual, by walking at random we mean that, upon reaching any vertex of the graph, the probability of choosing any one of the $2d$ edges leading out of that vertex is $1/2d$. We will call this random walk **simple random walk** in d dimensions.

When $d = 1$, our lattice is just an infinite line divided into segments of length one. We may think of the vertices of this graph as representing the fortune of a gambler betting on heads or tails in a fair coin tossing game. Simple random walk in one dimension then represents the vicissitudes of his or her fortune, either increasing or decreasing by one unit after each round of the game.

When $d = 2$, our lattice looks like an infinite network of streets and avenues, which is why we describe the random motion of the wandering point as a "walk."

When $d = 3$, the lattice looks like an infinite "jungle gym," so perhaps in this case we ought to talk about a "random climb," but we will not do so. It is worth noting that when $d = 3$, the wanderings of our point could be regarded as an approximate representation of the random path of a molecule diffusing in an infinite cubical crystal. Figure 5.2 shows a simulation of a simple random walk in three dimensions.

5.2 The Question of Recurrence

The question that Pólya posed amounts to this: "Is the wandering point certain to return to its starting point

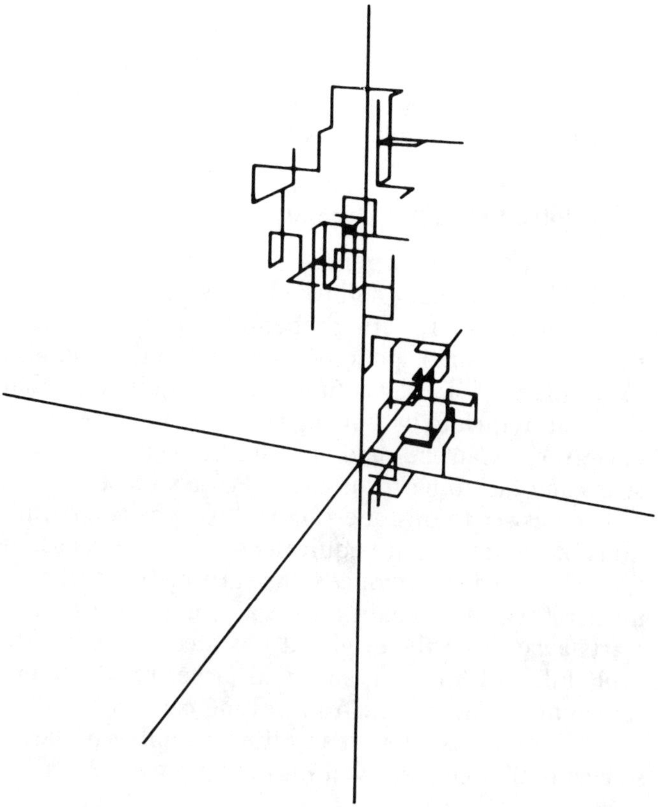

FIGURE 5.2

during the course of its wanderings?" If so, we say that the walk is **recurrent**. If not, that is, if there is a positive probability that the point will never return to its starting point, then we say that the walk is **transient**.

If we denote the probability that the point never returns to its starting point by p_{escape}, then the chain is

$$\textit{recurrent} \text{ if } p_{\text{escape}} = 0,$$

$$\textit{transient} \text{ if } p_{\text{escape}} > 0.$$

We will call the problem of determining recurrence or transience of a random walk the **type problem.**

5.3 Pólya's Original Question

The definition of recurrence that we have given differs from Pólya's original definition. Pólya defined a walk to be recurrent if, with probability one, it will pass through every single point of the lattice in the course of its wanderings. In our definition, we require only that the point return to its starting point. So we have to ask ourselves, "Can the random walk be recurrent in our sense and fail to be recurrent in Pólya's sense?"

The answer to this question is, "No, the two definitions of recurrence are equivalent." Why? Because if the point must return once to its starting point, then it must return there again and again, and each time it starts away from the origin, it has a certain non-zero probability of hitting a specified target vertex before returning to the origin. And anyone can get a bull's-eye if he or she is allowed an infinite number of darts, so eventually the point will hit the target vertex.

Exercise

1. Write out a rigorous version of the argument just given.

5.4 Pólya's Theorem: Recurrence in the Plane, Transience in Space

In the paper cited above, Pólya proved the following theorem:

PÓLYA'S THEOREM. *Simple random walk on a d-dimensional lattice is recurrent for $d = 1, 2$ and transient for $d > 2$.*

The rest of this book will be devoted to trying to understand this theorem. Our approach will be to exploit once more the connections between questions about a random walk on a graph and questions about electric currents in a corresponding network of resistors. We hope that this approach, by calling on methods that appeal to our physical intuition, will leave us feeling that we know "why" the theorem is true.

EXERCISES

1. Show that Pólya's theorem implies that if two random walkers start at $\mathbf{0}$ and wander independently, then in one and two dimensions they will eventually meet again, but in three dimensions there is positive probability that they won't.

2. Show that Pólya's theorem implies that a random walker in three dimensions will eventually hit the line defined by $x = 2, z = 0$.

5.5 The Escape Probability as a Limit of Escape Probabilities for Finite Graphs

We can determine the type of an infinite lattice from properties of bigger and bigger finite graphs that sit inside it. The simplest way to go about this is to look at the lattice analog of balls (solid spheres) in space. These are defined as follows: Let r be an integer—this will be the radius of the ball. Let $G^{(r)}$ be the graph gotten from $\mathbf{Z}^d$ by throwing out vertices whose distance from the origin is $>r$. By "distance from the origin" we mean here not the usual Euclidean distance, but the distance "in the lattice;" that is, the length of the shortest path along the edges of the lattice between the two points. Let $S^{(r)}$ be the "sphere" of radius r about the origin, i.e., those points that are exactly r units from the origin. In two dimensions, $S^{(r)}$ looks like a square. In three dimensions, it looks like an octahedron.

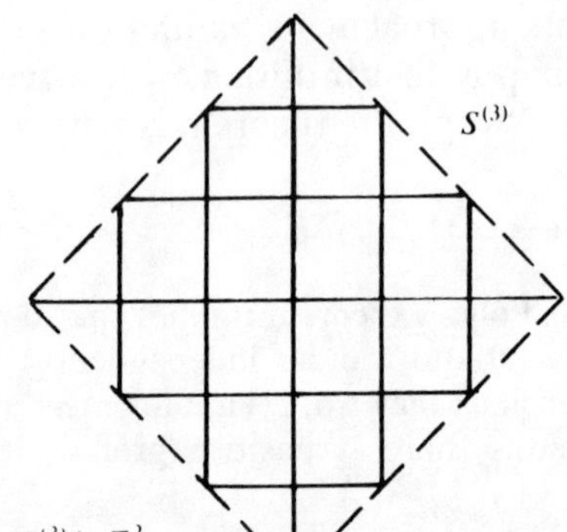

FIGURE 5.3

We define a random walk on $G^{(r)}$ as follows: The walk starts at $\mathbf{0}$ and continues as it would on $\mathbf{Z}^d$ until it reaches a point on $S^{(r)}$ and then it stays at this point. Thus the walk on $G^{(r)}$ is an absorbing Markov chain with every point of $S^{(r)}$ an absorbing state.

Let $p^{(r)}_{\text{escape}}$ be the probability that a random walk on $G^{(r)}$, starting at $\mathbf{0}$, reaches $S^{(r)}$ before returning to $\mathbf{0}$. Then $p^{(r)}_{\text{escape}}$ decreases as r increases and $p_{\text{escape}} = \lim_{r\to\infty} p^{(r)}_{\text{escape}}$ is the **escape probability** for the infinite graph. If this limit is 0, the infinite walk is recurrent. If it is greater than 0, the walk is transient.

5.6 Electrical Formulation of the Type Problem

Now that we have expressed things in terms of finite graphs, we can make use of the electrical methods introduced in Part I. To determine $p^{(r)}_{\text{escape}}$ electrically, we simply ground all the points of $S^{(r)}$, maintain **0** at one volt, and measure the current $i^{(r)}$ flowing into the circuit.

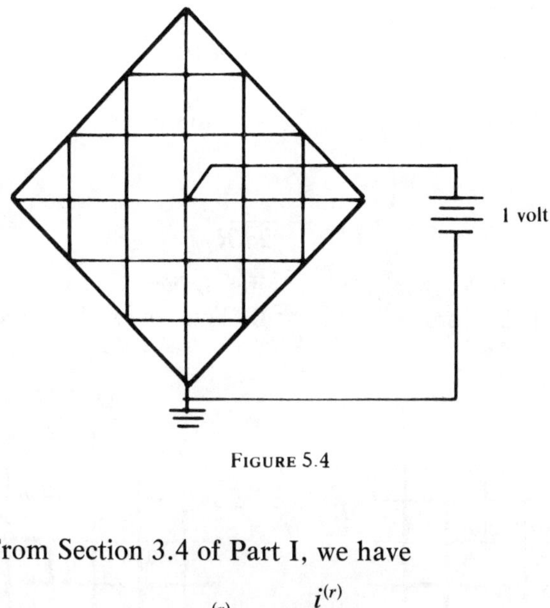

FIGURE 5.4

From Section 3.4 of Part I, we have

$$p^{(r)}_{\text{escape}} = \frac{i^{(r)}}{2d},$$

where d is the dimension of the lattice. (Remember that we have to divide by the number of branches coming out of the starting point.) Since the voltage being applied is 1, $i^{(r)}$ is just the effective conductance between **0** and $S^{(r)}$, i.e.,

$$i^{(r)} = \frac{1}{R_{EFF}^{(r)}},$$

where $R_{EFF}^{(r)}$ is the effective resistance from $\mathbf{0}$ to $S^{(r)}$. Thus

$$p_{\text{escape}}^{(r)} = \frac{1}{2d\,R_{EFF}^{(r)}}.$$

Define R_{EFF}, the **effective resistance from zero to infinity**, to be

$$R_{EFF} = \lim_{r\to\infty} R_{EFF}^{(r)}$$

This limit exists since $R_{EFF}^{(r)}$ is an increasing function of r. Then

$$p_{\text{escape}} = \frac{1}{2d\,R_{EFF}}.$$

Of course R_{EFF} may be infinite; in fact, this will be the case if and only if $p_{\text{escape}} = 0$. Thus the walk is recur-

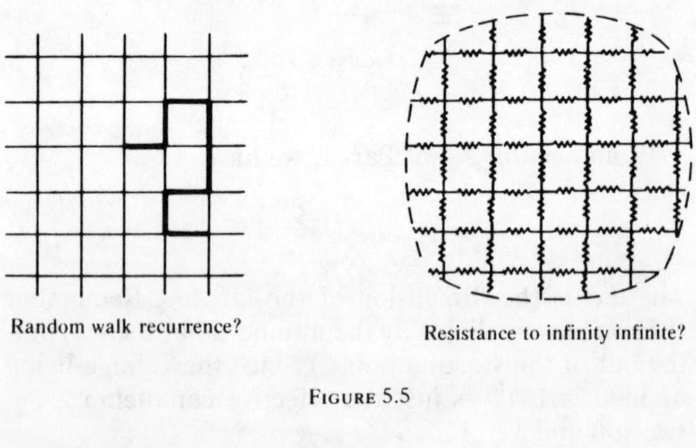

Random walk recurrence?

Resistance to infinity infinite?

FIGURE 5.5

rent if and only if the resistance to infinity is infinite, which makes sense.

The success of this electrical formulation of the type problem will come from the fact that the resistance to infinity can be estimated using classical methods of electrical theory.

5.7 One Dimension is Easy, but what about Higher Dimensions?

We now know that simple random walk on a graph is recurrent if and only if a corresponding network of 1-ohm resistors has infinite resistance "out to infinity." Since an infinite line of resistors obviously has infinite resistance, it follows that simple random walk on the 1-dimensional lattice is recurrent, as stated by Pólya's theorem.

Resistance to infinity infinite!

FIGURE 5.6

What happens in higher dimensions? We are asked to decide whether a d-dimensional lattice has infinite resistance to infinity. The difficulty is that the d-dimensional lattice $\mathbf{Z}^d$ lacks the rotational symmetry of the Euclidean space $\mathbf{R}^d$ in which it sits.

To see how this lack of symmetry complicates electrical problems, we determine, by solving the appropriate discrete Dirichlet problem, the voltages for a one-volt battery attached between $\mathbf{0}$ and the points of $S^{(3)}$ in $\mathbf{Z}^2$. The resulting voltages are:

			0			
		0	.091	0		
	0	.182	.364	.182	0	
0	.091	.364	1	.364	.091	0
	0	.182	.364	.182	0	
		0	.091	0		
			0			

The voltages at points of $S^{(1)}$ are equal, but the voltages at points of $S^{(2)}$ are not. This means that the resistance from $\mathbf{0}$ to $S^{(3)}$ cannot be written simply as the sum of the resistances from $\mathbf{0}$ to $S^{(1)}$, $S^{(1)}$ to $S^{(2)}$, and $S^{(2)}$ to $S^{(3)}$. This is in marked contrast to the case of a continuous resistive medium to be discussed in the next section.

Exercises

1. Using the voltages given for $G^{(3)}$, find $R^{(3)}_{EFF}$ and $p^{(3)}_{\text{escape}}$.

2. Consider a one-dimensional infinite network with resistors $R_n = 1/2^n$ from n to $n + 1$ for $n = 0, \pm 1, \pm 2, \ldots$. Describe the associated random walk and determine whether the walk is recurrent or transient.

3. A random walk moves on the non-negative integers; when it is in state n, $n > 0$, it moves with probability p_n to $n + 1$ and with probability $1 - p_n$ to $n - 1$. When at 0, it moves to 1. Determine a network that gives this random walk and give a criterion in terms of the p_n for recurrence of the random walk.

5.8 Getting Around the Lack of Rotational Symmetry of the Lattice

Suppose we replace our d-dimensional resistor lattice by a (homogeneous, isotropic) resistive medium filling all of $\mathbf{R}^d$ and ask for the effective resistance to infinity. Naturally we expect that the rotational symmetry will make this continuous problem easier to solve than the original discrete problem. If we took this problem to a physicist, he or she would probably produce something like the scribblings illustrated in Figure 5.7 and conclude that the effective resistance is infinite for $d = 1$, 2 and finite for $d > 2$. The analogy to Pólya's theorem is obvious, but is it possible to translate these calculations for continuous media into information about what happens in the lattice?

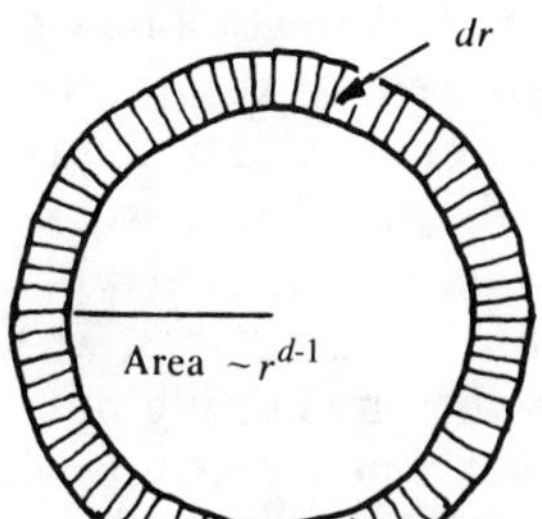

$$R_\infty \sim \int_a^\infty \frac{dr}{r^{d-1}} \quad \begin{matrix} = \infty & \text{for} \quad d = 1, 2 \\ < \infty & \text{for} \quad d < 3 \end{matrix}$$

FIGURE 5.7

This can indeed be done, and this would certainly be the most natural approach to take. We will come back to this approach at the end of the book. For now, we will take a different approach to getting around the asymmetry of the lattice. Our method will be to modify the lattice in such a way as to obtain a graph that is symmetrical enough so that we can calculate its resistance out to infinity. Of course, we will have to think carefully about what happens to that resistance when we make these modifications.

5.9 Rayleigh: Shorting Shows Recurrence in the Plane, Cutting Shows Transience in Space

Here is a sketch of the method we will use to prove Pólya's theorem. The details will be given in Chapter 6.

To take care of the case $d = 2$, we will modify the 2-dimensional resistor network by shorting certain sets of nodes together so as to get a new network whose resistance is readily seen to be infinite. As shorting can only decrease the effective resistance of the network, the resistance of the original network must also be infinite. Thus the walk is recurrent when $d = 2$.

To take care of the case $d = 3$, we will modify the 3-dimensional network by cutting out certain of the resistors so as to get a new network whose resistance is readily seen to be finite. As cutting can only increase the resistance of the network, the resistance of the original network must also be finite. Thus the walk is transient when $d = 3$.

The method of applying shorting and cutting to get lower and upper bounds for the resistance of a resistive medium was introduced by Lord Raylcigh. (Rayleigh [**31**]; see also Maxwell [**23**], Jeans [**12**], Pólya and Szegö [**30**]). We will refer to Rayleigh's techniques collectively as **Rayleigh's short-cut method**. This does not do Rayleigh justice, for Rayleigh's method is a whole bag of tricks that goes beyond mere shorting and cutting—but who can resist a pun?

Rayleigh's method was apparently first applied to random walks by C. St. J. A. Nash-Williams [**26**], who used the shorting method to establish recurrence for random walks on the 2-dimensional lattice.

CHAPTER 6

RAYLEIGH'S SHORT-CUT METHOD

6.1 Shorting and Cutting

In its simplest form, Rayleigh's method involves modifying the network whose resistance we are interested in so as to get a simpler network. We consider two kinds of modifications, shorting and cutting. Cutting involves nothing more than clipping some of the branches of the network, or what is the same, simply deleting them from the network. Shorting involves connecting a given set of nodes together with perfectly conducting wires, so that current can pass freely be-

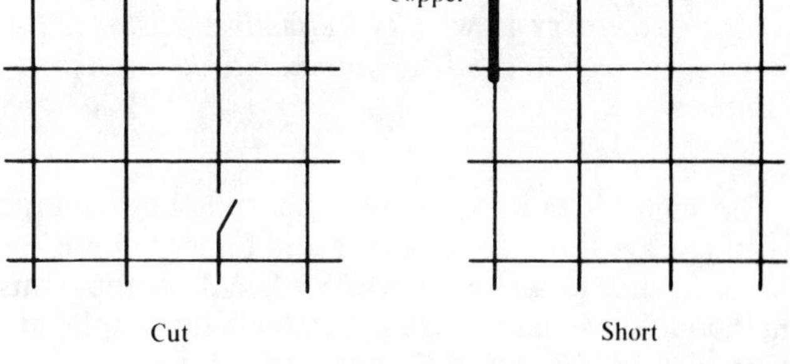

FIGURE 6.1

tween them. In the resulting network, the nodes that were shorted together behave as if they were a single node.

6.2 The Shorting Law and the Cutting Law; Rayleigh's Idea

The usefulness of these two procedures (shorting and cutting) stems from the following observations:

SHORTING LAW. *Shorting certain sets of nodes together can only decrease the effective resistance of the network between two given nodes.*

CUTTING LAW. *Cutting certain branches can only increase the effective resistance between two given nodes.*

These laws are both equivalent to Rayleigh's Monotonicity Law, which was introduced in Chapter 4 of Part I (see Exercise 1):

MONOTONICITY LAW. *The effective resistance between two given nodes is monotonic in the branch resistances.*

Rayleigh's idea was to use the Shorting Law and the Cutting Law above to get lower and upper bounds for the resistance of a network. In Section 3 we apply this method to solve the recurrence problem for simple random walk in dimensions 2 and 3.

EXERCISE

1. Show that the Shorting Law and the Cutting Law are both equivalent to the Monotonicity Law.

6.3 The Plane is Easy

When $d = 2$, we apply the Shorting Law as follows: Short together nodes on squares about the origin, as shown in Figure 6.2. The network we obtain is equivalent to the network shown in Figure 6.3.

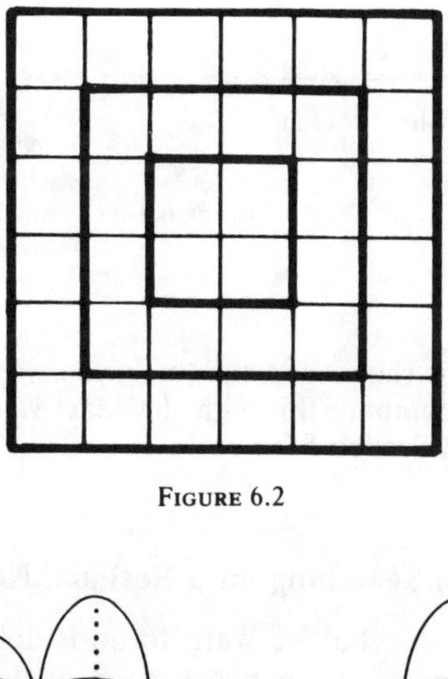

FIGURE 6.2

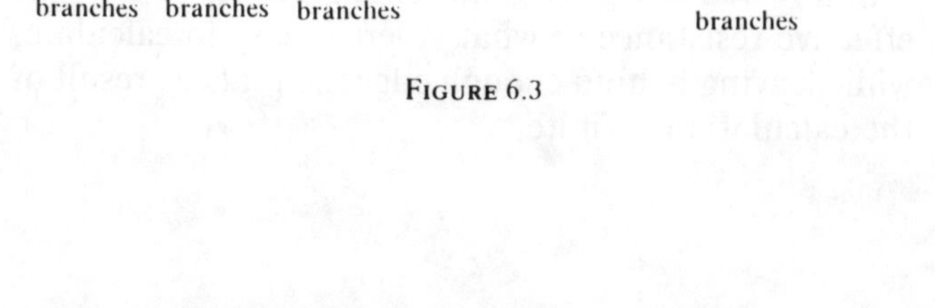

FIGURE 6.3

Now as n 1-ohm resistors in parallel are equivalent to a single resistor of resistance $1/n$ ohms, the modified network is equivalent to the network shown in Figure 6.4. The resistance of this network out to infinity is

$$\sum_{n=0}^{\infty} \frac{1}{8n + 4} = \infty.$$

As the resistance of the old network can only be bigger, we conclude that it too must be infinite, so that the walk is recurrent when $d = 2$.

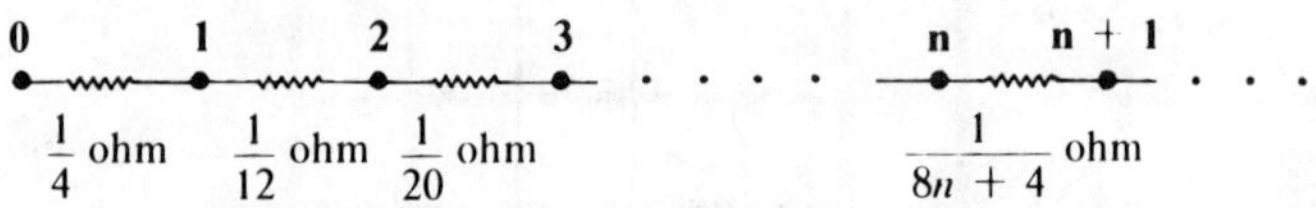

FIGURE 6.4

EXERCISE

1. Using the shorting technique, give an upper bound for $p^{(3)}_{\text{escape}}$ and compare this with the exact value obtained in Exercise 1 of Section 5.7.

6.4 Space: Searching for a Residual Network

When $d = 3$, what we want to do is delete certain of the branches of the network so as to leave behind a residual network having manifestly finite resistance. The problem is to reconcile the "manifestly" with the "finite." We want to cut out enough edges so that the effective resistance of what is left is easy to calculate, while leaving behind enough edges so that the result of the calculation is finite.

6.5 Trees are Easy to Analyze

Trees—that is, graphs without circuits—are undoubtedly the easiest to work with. For instance, consider the full binary tree, shown in Figure 6.5. Notice that sitting inside this tree just above the root are two copies

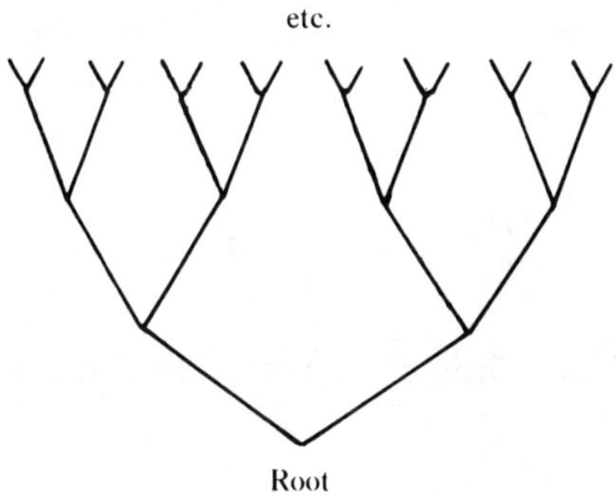

The full binary tree

FIGURE 6.5

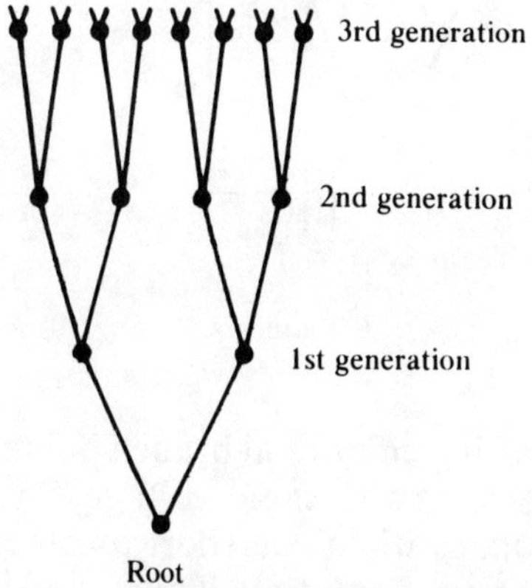

FIGURE 6.6

of the tree itself. This self-similarity property can be used to compute the effective resistance R_∞ from the root out to infinity. (See Exercise 1.) It turns out that $R_\infty = 1$. We will demonstrate this below by a more direct method.

To begin with, let us determine the effective resistance R_n between the root and the set of nth generation branch points. To do this, we should ground the set of branch points, hook the root up to a 1-volt battery, and compute

$$R_n = \frac{1}{\text{current through battery}}.$$

For $n = 3$, the circuit that we would obtain is shown in Figure 6.7.

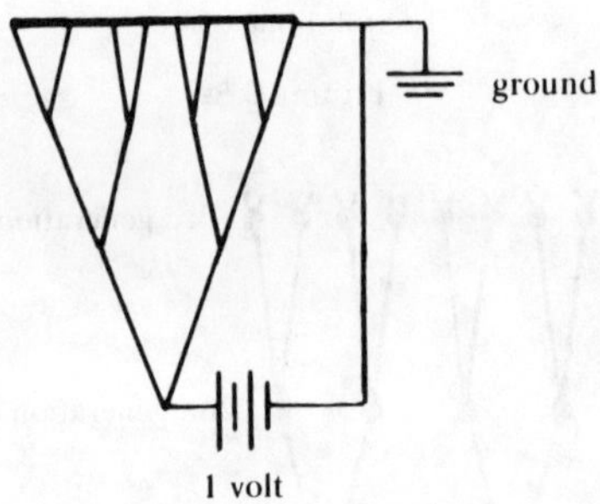

FIGURE 6.7

In the resulting circuit, all branch points of the same generation are at the same voltage (by symmetry). Nothing happens when you short together nodes that are at the same potential. Thus shorting together branch points of the same generation will not affect the distribution of currents in the branches. In particular, this modification will not affect the current through the battery, and we conclude that

$$R_n = \frac{1}{\begin{array}{c}\text{current in original}\\ \text{circuit}\end{array}} = \frac{1}{\begin{array}{c}\text{current in modified}\\ \text{circuit}\end{array}}.$$

For $n = 3$, the modified circuit is shown in Figure 6.8. This picture shows that $R_3 = 1/2 + 1/4 + 1/8 = 1 - 1/2^3$. More generally,

$$R_n = \frac{1}{2} + \frac{1}{4} + \cdots + \frac{1}{2^n} = 1 - \frac{1}{2^n}.$$

Letting $n \to \infty$, we get

$$R_\infty = \lim_{n\to\infty} R_n = \lim_{n\to\infty} 1 - \frac{1}{2^n} = 1.$$

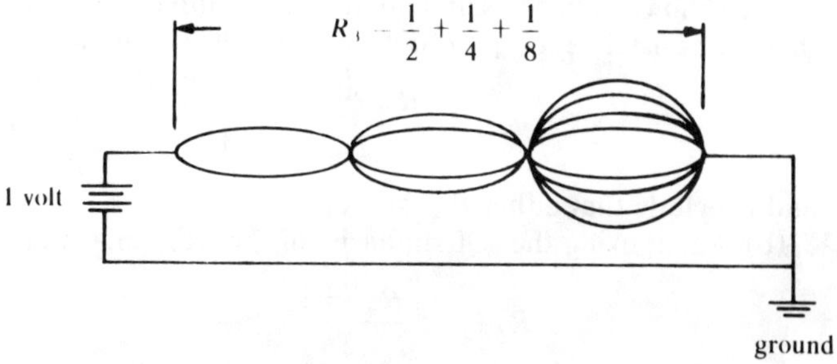

FIGURE 6.8

Figure 6.9 shows another closely related tree, the **tree homogeneous of degree three**: Note that all nodes of this tree are similar—there is no intrinsic way to distinguish one from another. This tree is obviously a

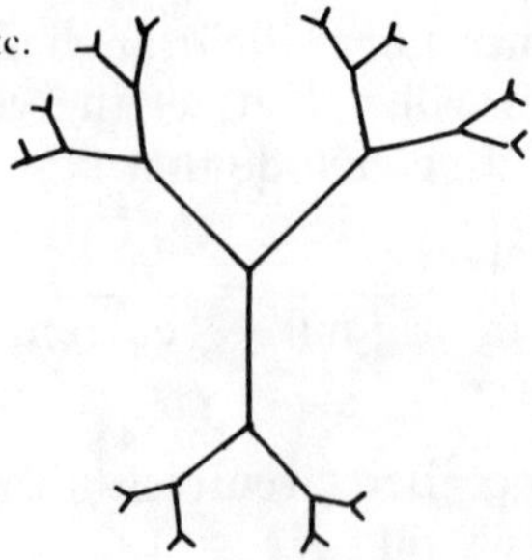

The tree homogeneous of degree 3

FIGURE 6.9

close relative of the full binary tree. Its resistance to infinity is 2/3.

EXERCISE

1. (a) Show, using the self-similarity of the full binary tree, that the resistance R_∞ to infinity satisfies the equation

$$R_\infty = \frac{R_\infty + 1}{2}$$

and conclude that either $R_\infty = 1$ or $R_\infty = \infty$.

(b) Again using the self-similarity of the tree, show that

$$R_{n+1} = \frac{R_n + 1}{2}$$

where R_n denotes the resistance out to the set of the nth generation branch points. Conclude that $R_\infty = \lim_{n\to\infty} R_n = 1$.

6.6 The Full Binary Tree is Too Big

Nothing could be nicer than the two trees we have just described. They are the prototypes of networks having manifestly finite resistance to infinity. Unfortunately, we can't even come close to finding either of these trees as a subgraph of the three-dimensional lattice. For in these trees, the number of nodes in a "ball" of radius r grows exponentially with r, whereas in a d-dimensional lattice, it grows like r^d, i.e., much slower. (See Figure 6.10.) There is simply no room for these trees in any finite-dimensional lattice.

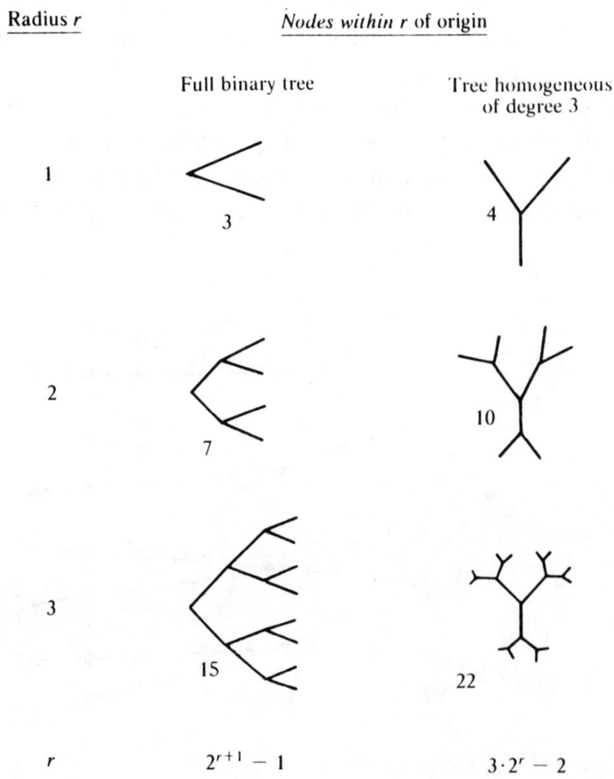

FIGURE 6.10

6.7 NT_3: A "Three-Dimensional" Tree

These observations suggest that we would do well to look for a nice tree NT_3 where the number of nodes within a radius r of the root is on the order of r^3. For we might hope to find something resembling NT_3 in the 3-dimensional lattice, and if there is any justice in the world, this tree would have finite resistance to infinity, and we would be done.

Before introducing NT_3, let's have a look at NT_2, our choice for the tree most likely to succeed in the 2-dimensional lattice (see Figure 6.11). The idea behind NT_2 is that, since a ball of radius r in the graph ought to contain something like r^2 points, a sphere of radius r ought to contain something like r points, so the number of points in a sphere should roughly dou-

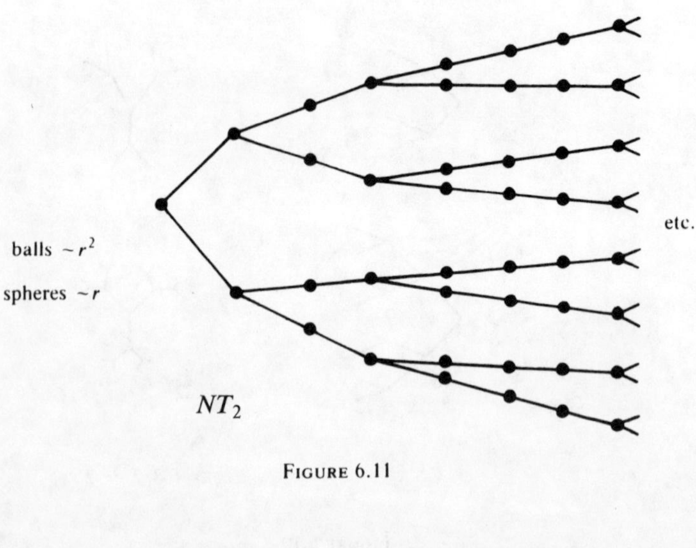

FIGURE 6.11

ble when the radius of the sphere is doubled. For this reason, we make the branches of our tree split in two every time the distance from the origin is (roughly) doubled.

Similarly, in a 3-dimensional tree, when we double the radius, the size of a sphere should roughly quadruple. Thus in NT_3, we make the branches of our tree split in four where the branches of NT_2 would have split in two. NT_3 is shown in Figure 6.12. Obviously, NT_3 is none to happy about being drawn in the plane.

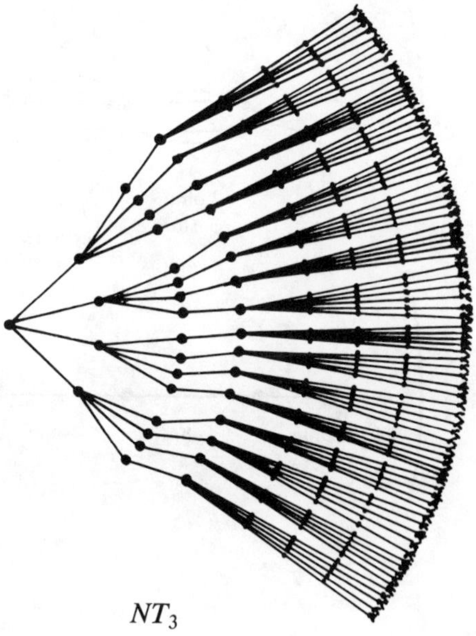

NT_3

FIGURE 6.12

6.8 NT_3 has Finite Resistance

To see if we're on the right track, let's work out the resistance of our new trees. These calculations are shown in Figures 6.13 and 6.14.

$R_{\text{root},\infty}(NT_2) = R_{\text{root},\infty}$

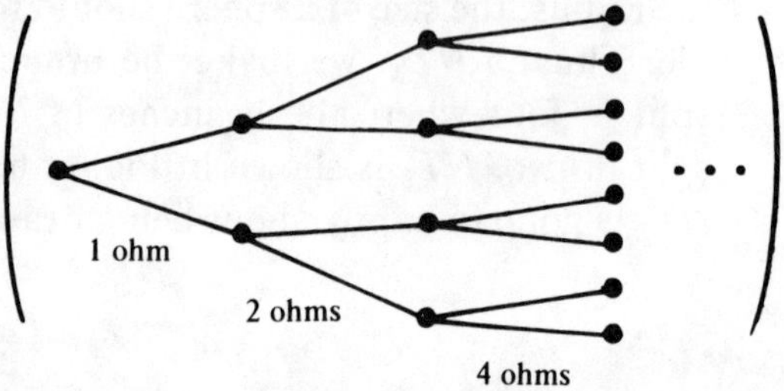

$= R_{\text{root},\infty}$

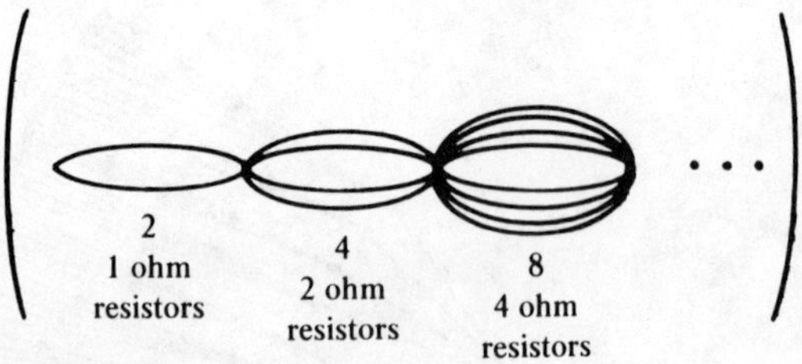

$= R_{\text{root},\infty}$

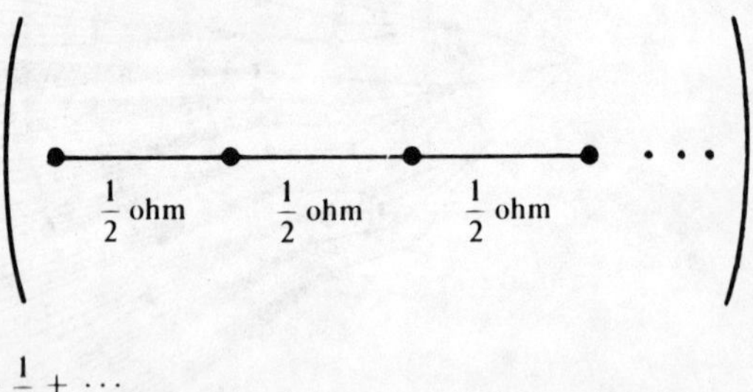

$$= \frac{1}{2} + \frac{1}{2} + \frac{1}{2} + \cdots$$

$$= \infty \, .$$

FIGURE 6.13

$$R_{\text{root},\infty}(NT_3) = R_{\text{root},\infty}\Bigg($$

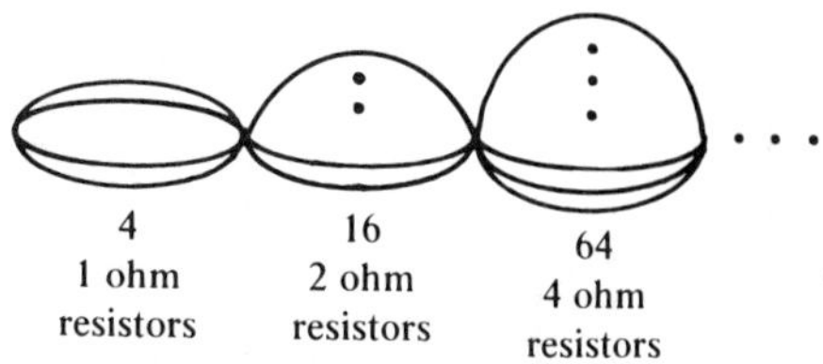

$$\Bigg)$$

$$= R_{\text{root},\infty}\Bigg($$

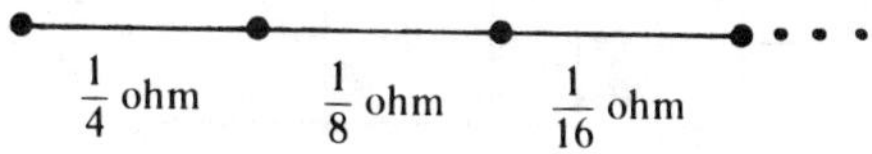

$$\Bigg)$$

$$= \frac{1}{4} + \frac{1}{8} + \frac{1}{16} + \cdots$$

$$= \frac{1}{2}\,.$$

FIGURE 6.14

As we would hope, the resistance of NT_2 is infinite, but the resistance of NT_3 is not.

EXERCISE

1. Use self-similarity arguments to compute the resistance of NT_2 and NT_3.

6.9 But Does NT$_3$ Fit in the Three-Dimensional Lattice?

We would like to embed NT_3 in $\mathbf{Z}^3$. We start by trying to embed NT_2 in $\mathbf{Z}^2$. The result is shown in Figure 6.15.

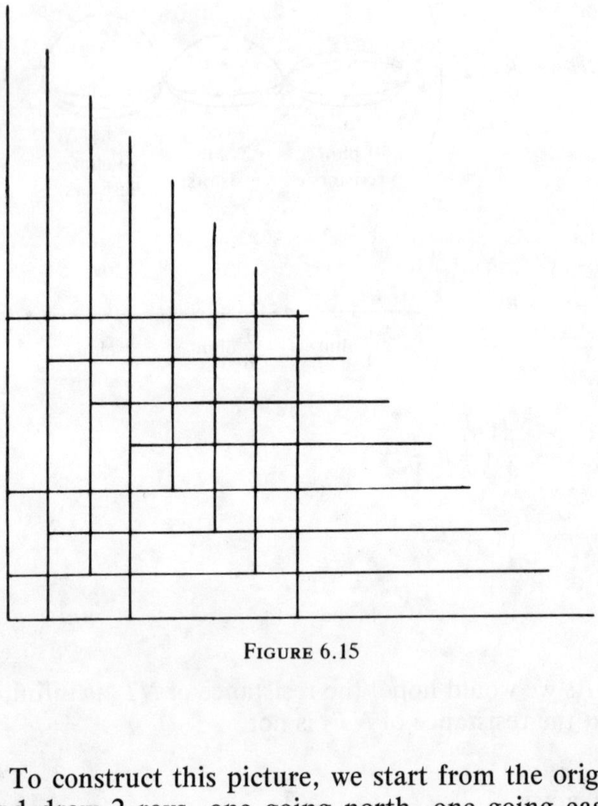

FIGURE 6.15

To construct this picture, we start from the origin and draw 2 rays, one going north, one going east. Whenever a ray intersects the line $x + y = 2^n - 1$ for some n, it splits into 2 rays, one going north, and one going east. The sequence of pictures in Figure 6.16 shows successively larger portions of the graph, along with the corresponding portions of NT_2.

Of course this isn't really an embedding, since certain pairs of points that were distinct in NT_2 get identified, that is, they are made to correspond to a single point in the picture. In terms of our description, sometimes a ray going north and a ray going east pass through each other. This could have been avoided by allowing the rays to "bounce" instead of passing through each other, at the expense of embedding not NT_2 but a close relative—see Exercise 3. However, because the points of each identified pair are at the same distance from the root of NT_2, when we put a battery between the root and the nth level they will be at the same potential. Hence, the current flow is not affected by these identifications, so the identifications have no effect on R_{EFF}. For our purposes, then, we have done just as well as if we had actually embedded NT_2.

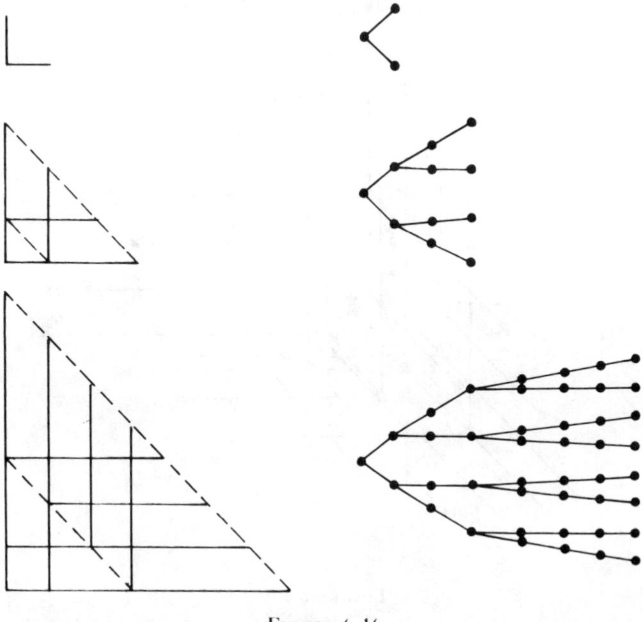

FIGURE 6.16

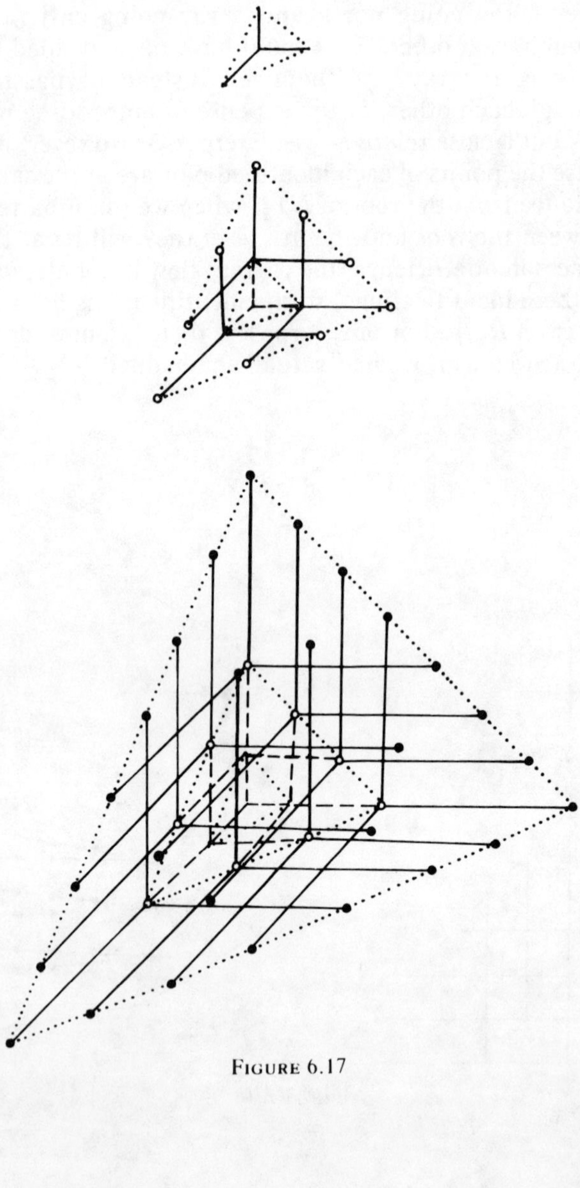

FIGURE 6.17

To construct the analogous picture in three dimensions, we start three rays off from the origin going north, east, and up. Whenever a ray intersects the plane $x + y + z = 2^n - 1$ for some n, it splits into three rays, going north, east, and up. This process is illustrated in Figure 6.17.

Surprisingly, the subgraph of the 3-dimensional lattice obtained in this way is not NT_3! Rather, it represents an attempt to embed the tree shown in Figure 6.18.

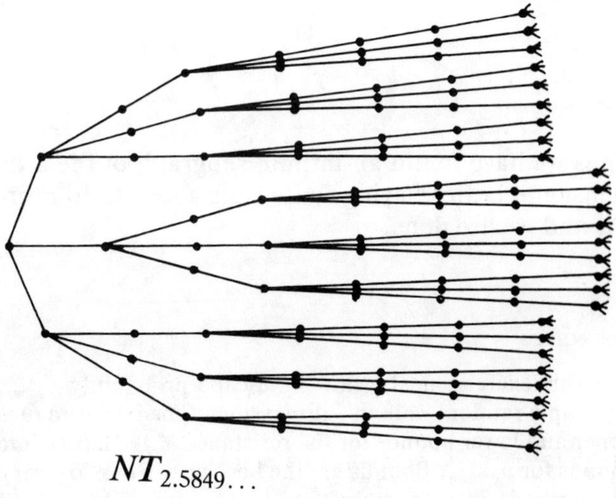

$NT_{2.5849\ldots}$

FIGURE 6.18

We call this tree $NT_{2.5849\ldots}$ because it is 2.5849...-dimensional in the sense that when you double the radius of a ball, the number of points in the ball gets multiplied roughly by 6 and

$$6 = 2^{\log_2 6} = 2^{2.5849\ldots}.$$

Again, certain pairs of points of $NT_{2.5849\ldots}$ have been allowed to correspond to the same point in the lattice, but once again the intersections have no effect on R_{EFF}.

So we haven't come up with our embedded NT_3 yet. But why bother? The resistance of $NT_{2.5849\ldots}$ out to infinity is

$$\frac{1}{3} + \frac{2}{9} + \frac{4}{27} + \cdots = \frac{1}{3}\left(1 + \frac{2}{3} + \left(\frac{2}{3}\right)^2 + \cdots\right)$$

$$= \frac{1}{3}\,\frac{1}{1 - \frac{2}{3}} = 1.$$

Thus we have found an infinite subgraph of the 3-dimensional lattice having finite resistance out to infinity, and we are done.

Exercises

1. This exercise deals with the escape probability p_{escape} for simple random walk in 3 dimensions. The idea is to turn upper and lower bounds for the resistance of the lattice into bounds for p_{escape}. Bounds are the best we can ask for using our method. The determination of the exact value will be discussed in Section 7.5. It is roughly .66.

(a) Use a shorting argument to get an upper bound for p_{escape}.

(b) We have seen that the resistance of the 3-dimensional lattice is at most one ohm. Show that the corresponding lower bound for p_{escape} is 1/6. Show that this lower bound can be increased to 1/3 with no extra effort.

2. Prove that simple random walk in any dimension $d > 3$ is transient.

3. Show how the not-quite embeddings of NT_2 and $NT_{2.5849\ldots}$ can be altered to yield honest-to-goodness embeddings of "stretched-out" versions of these trees, obtained by replacing each edge of the tree by 3 edges in series. (Hint: "bounce.")

6.10 What We Have Done; What We Will Do

We have finally finished our electrical proof of Pólya's theorem. The proof of recurrence for $d = 1, 2$ was straight-forward, but this could hardly be said of the proof for $d = 3$. After all, we started out trying to embed NT_3 and ended up by not quite embedding something that was not quite NT_3!

This is not bad in itself, for one frequently sets out to do something and in the process of trying to do it gets a better idea. The real problem is that this explicit construction is just too clever, too artificial. We seem to be saying that a simple random walk in 3 dimensions is transient because $\mathbf{Z}^3$ happens to contain a beautifully symmetrical subgraph that is in some sense 2.5849...-dimensional! Fine, but what if we hadn't stumbled upon this subgraph? Isn't there some other, more natural way?

We will see that indeed there are more natural approaches to showing transience for $d = 3$. One such approach uses the same idea of embedding trees, but depends on the observation that one doesn't need to be too careful about sending edges to edges. Another approach, based on relating the lattice not to a tree but to Euclidean space, was already hinted at in Section 5.8. The main goal for the rest of this book will be to explore these more natural electrical approaches to Pólya's theorem.

Before jumping into this, however, we are going to go back and take a look at a classical—i.e., probabilistic—approach to Pólya's theorem. This will give us something to compare our electrical proofs with.

CHAPTER 7

THE CLASSICAL PROOFS OF PÓLYA'S THEOREM

7.1 Recurrence is Equivalent to an Infinite Expected Number of Returns

In this chapter, all of our random walks will be simple. Let u be the probability that a random walker, starting at the origin, will return to the origin. The probability that the walker will be there exactly k times (counting the initial time) is $u^{k-1}(1 - u)$. Thus, if m is the expected number of times at the origin,

$$m = \sum_{k=1}^{\infty} ku^{k-1}(1 - u)$$

$$= \frac{1}{1 - u}.$$

If $m = \infty$ then $u = 1$, and hence the walk is recurrent. If $m < \infty$ then $u < 1$, so the walk is transient. Thus m determines the type of the walk.

We shall use an alternate expression for m. Let u_n be the probability that the walk, starting at $\mathbf{0}$, is at $\mathbf{0}$ on the nth step. Since the walker starts at $\mathbf{0}$, $u_0 = 1$. Let e_n be a random variable that takes on the value 1 if, at time n, the walker is at $\mathbf{0}$ and 0 otherwise. Then

$$T = \sum_{n=0}^{\infty} e_n$$

is the total number of times at $\mathbf{0}$ and

$$m = E(T) = \sum_{n=0}^{\infty} E(e_n).$$

But $E(e_n) = 1 \cdot u_n + 0 \cdot (1 - u_n) = u_n$. Thus

$$m = \sum_{n=0}^{\infty} u_n .$$

Therefore, the walk will be recurrent if the series $\Sigma_{n=0}^{\infty} u_n$ diverges and transient if it converges.

EXERCISE

1. Let $N_{\mathbf{xy}}$ be the expected number of visits to $\mathbf{y}$ for a random walker starting in $\mathbf{x}$. Show that $N_{\mathbf{xy}}$ is finite if and only if the walk is transient.

7.2 Simple Random Walk in One Dimension

Consider a random walker in one dimension, started at $\mathbf{0}$. To return to $\mathbf{0}$, the walker must take the same number of steps to the right as to the left; hence, only even times are possible. Let us compute u_{2n}. Any path that returns in $2n$ steps has probability $1/2^{2n}$. The number of possible paths equals the number of ways that we can choose the n times to go right from the $2n$ possible times. Thus

$$u_{2n} = \binom{2n}{n} \frac{1}{2^{2n}}.$$

We shall show that $\Sigma_n u_{2n} = \infty$ by using Stirling's approximation:

$$n! \sim \sqrt{2\pi n}\, e^{-n} n^n.$$

Thus

$$u_{2n} = \frac{(2n)!}{n!n!}\frac{1}{2^{2n}} \sim \frac{\sqrt{2\pi \cdot 2n}\, e^{-2n}(2n)^{2n}}{(\sqrt{2\pi n}\, e^{-n} n^n)^2\, 2^{2n}}$$

$$= \frac{1}{\sqrt{\pi n}}.$$

Therefore,

$$\sum_n u_{2n} \sim \sum_n \frac{1}{\sqrt{\pi n}} = \infty$$

and a simple random walk in one dimension is recurrent.

Recall that this case was trivial by the resistor argument.

Exercises

1. We showed in Part I that a random walker starting at x with $0 < x < N$ has probability x/N of reaching N before 0. Use this to show that a simple random walk in one dimension is recurrent.

2. Consider a random walk in one dimension that moves from n to $n + 1$ with probability p and to $n - 1$ with probability $q = 1 - p$. Assume that $p > 1/2$. Let h_x be the probability, starting at x, that the walker ever reaches 0. Use Exercise 1 of Section 1.6 in Part I to show that $h_x = (q/p)^x$ for $x \geqslant 0$ and $=1$ for $x < 0$. Show that this walk is transient.

3. For a simple random walk in one dimension, it follows from Exercise 4 of Section 1.5 in Part I that the expected

time, for a walker starting at x with $0 < x < N$, to reach 0 or N is $x(N - x)$. Prove that for the infinite walk, the expected time to return to 0 is infinite.

4. Let us regard a simple random walk in one dimension as the fortune of a player in a penny matching game where the players have unlimited credit. Show that the result is a martingale (see Part I, Section 1.6). Show that you can describe a stopping system that guarantees that you make money.

7.3 Simple Random Walk in Two Dimensions

For a random walker in two dimensions to return to the origin, the walker must have gone the same number of times north and south and the same number of times east and west. Hence, again, only even times for return are possible. Every path that returns in $2n$ steps has probability $1/4^{2n}$. The number of paths that do this by taking k steps to the north, k south, $n - k$ east and $n - k$ west is

$$\binom{2n}{k, k, n-k, n-k} = \frac{2n!}{k!k!(n-k)!(n-k)!}.$$

Thus

$$\begin{aligned} u_{2n} &= \frac{1}{4^{2n}} \sum_{k=0}^{n} \frac{(2n)!}{k!k!(n-k)!(n-k)!} \\ &= \frac{1}{4^{2n}} \sum_{k=0}^{n} \frac{(2n)!}{n!n!} \frac{n!n!}{k!k!(n-k)!(n-k)!} \\ &= \frac{1}{4^{2n}} \binom{2n}{n} \sum_{k=0}^{n} \binom{n}{k}^2. \end{aligned}$$

But $\Sigma_{k=0}^{n} \binom{n}{k}^2 = \binom{2n}{n}$ (see Exercise 1). Hence

$$u_{2n} = \left(\frac{1}{2^{2n}} \binom{2n}{n}\right)^2.$$

This is just the square of the one dimension result (not by accident, either—see Section 7.6). Thus we have

$$m = \sum_n u_{2n} \approx \sum_n \frac{1}{\pi n} = \infty.$$

Recall that the resistor argument was very simple in this case also.

EXERCISE

1. Show that $\Sigma_{k=0}^{n} \binom{n}{k}^2 = \binom{2n}{n}$. (Hint: Think of choosing n balls from a box that has $2n$ balls, n black and n white.)

7.4 Simple Random Walk in Three Dimensions

For a walker in three dimensions to return, the walker must take an equal number of steps in the three different directions. Thus we have

$$u_{2n} = \frac{1}{6^{2n}} \sum_{j,k} \frac{(2n)!}{j!j!k!k!(n-j-k)!(n-j-k)!}$$

where the sum is taken over all j, k with $j + k \leqslant n$. Following Feller [5], we rewrite this sum as

$$u_{2n} = \frac{1}{2^{2n}} \binom{2n}{n} \sum_{j,k} \left(\frac{1}{3^n} \frac{n!}{j!k!(n-j-k)!} \right)^2.$$

Now consider placing n balls randomly into three boxes, A, B, C. The probability that j balls fall in A, k in B, and $n - j - k$ in C is

$$\frac{1}{3^n} \binom{n}{j,\, k,\, n-j-k} = \frac{1}{3^n} \frac{n!}{j!k!(n-j-k)!}.$$

Intuitively, this probability is largest when j, k, and $n - j - k$ are as near as possible to $n/3$, and this can be proved (see Exercise 1). Hence, replacing one of the factors in the square by this larger value, we have:

$$u_{2n} \leqslant$$

$$\frac{1}{2^{2n}}\binom{2n}{n}\left(\frac{1}{3^n}\frac{n!}{\left[\frac{n}{3}\right]!\left[\frac{n}{3}\right]!\left[\frac{n}{3}\right]!}\right)\left(\sum_{j,k}\frac{1}{3^n}\frac{n!}{j!k!(n-j-k)!}\right).$$

Where $[n/3]$ denotes the greatest integer $\leqslant n/3$. The last sum is 1 since it is the sum of all probabilities for the outcomes of putting n balls into 3 boxes. Thus

$$u_{2n} \leqslant \frac{1}{2^{2n}}\binom{2n}{n}\left(\frac{1}{3^n}\frac{n!}{\left[\frac{n}{3}\right]!^3}\right).$$

Applying Stirling's approximation yields

$$u_{2n} \leqslant \frac{K}{n^{3/2}}$$

for suitable constant K. Therefore

$$m = \sum_n u_{2n} \leqslant K \sum_n \frac{1}{n^{3/2}} < \infty,$$

and a simple random walk in three dimensions is recurrent.

While this is a complex calculation, the resistor argument was also complicated in this case. We will try to make amends for this presently.

EXERCISES

1. Prove that $\binom{n}{j,k,n-j-k}$ is largest when j, k, and $n-j-k$ are as close as possible to $n/3$.

2. Find an appropriate value for the "suitable constant" K that was mentioned above, and derive an upper bound for m. Use this to get a lower bound for the probability of escape for simple random walk in three dimensions.

7.5 The Probability of Return in Three Dimensions: Exact Calculations

Since the probability of return in three dimensions is less than one, it is natural to ask, "What is this probability?" For this we need an exact calculation. The first such calculation was carried out in a pioneering paper of McCrea and Whipple [**24**]. The solution outlined here follows Feller [**5**], Exercise 28, Chapter 9, and Montroll and West [**25**].

Let $p(a, b, c; n)$ be the probability that a random walker, starting at 0, is at (a, b, c) after n steps. Then $p(a, b, c; n)$ is completely determined by the fact that

$$p(0, 0, 0; 0) = 1$$

and

$$p(a, b, c; n) =$$

$$\frac{1}{6}p(a-1, b, c; n-1) + \frac{1}{6}p(a+1, b, c; n-1)$$

$$+ \frac{1}{6}p(a, b-1, c; n-1) + \frac{1}{6}p(a, b+1, c; n-1)$$

$$+ \frac{1}{6}p(a, b, c-1; n-1) + \frac{1}{6}p(a, b, c+1; n-1).$$

Using the technique of generating functions, it is possible to derive a solution of these equations as

$$p(a, b, c; n) = \frac{1}{(2\pi)^3} \cdot \int_{-\pi}^{\pi}\int_{-\pi}^{\pi}\int_{-\pi}^{\pi} \left(\frac{\cos(x) + \cos(y) + \cos(z)}{3}\right)^n \cdot \cos(xa)\cos(yb)\cos(zc)dxdydz.$$

Of course, we can just verify that this formula satisfies our equations once someone has suggested it. Having this formula, we put $a = b = c = 0$ and sum over n to obtain the expected number of returns as

$$m = \frac{3}{(2\pi)^3} \cdot \int_{-\pi}^{\pi}\int_{-\pi}^{\pi}\int_{-\pi}^{\pi} \frac{1}{(3 - (\cos x + \cos y + \cos z))} dxdydz.$$

This integral was first evaluated by Watson [**37**] in terms of elliptic integrals, which are tabulated. A simpler result was obtained by Glasser and Zucker [**9**] who evaluated this integral in terms of gamma functions. Using this result, we get

$$m = \frac{\sqrt{6}}{32\pi^3}\Gamma\left(\frac{1}{24}\right)\Gamma\left(\frac{5}{24}\right)\Gamma\left(\frac{7}{24}\right)\Gamma\left(\frac{11}{24}\right) = 1.516386059137\ldots,$$

where $\Gamma(x)$ is Euler's gamma function, $\Gamma(x) = \int_0^\infty e^{-t}t^{x-1}dt$. (Incidentally, the value given by Glasser and Zucker [**9**] for the integral above needs to be corrected by multiplying by a factor of $1/(384\pi)$.)

Recall that $m = 1/(1 - u)$ so that $u = (m - 1)/m$. This gives

$$u = .340537329544\ldots.$$

7.6 Simple Random Walk in Two Dimensions is the Same as Two Independent One-Dimensional Random Walks

We observed that the probability of return at time $2n$ in two dimensions is the square of the corresponding probability in one dimension. Thus it is the same as the probability that two independent walkers, one walking in the x direction and the other in the y direction, will, at time $2n$, both be at (0, 0). Can we see that this should be the case? The answer is yes. Just change our axes by 45 degrees to new axes $\bar{x}$ and $\bar{y}$ as in Figure 7.1.

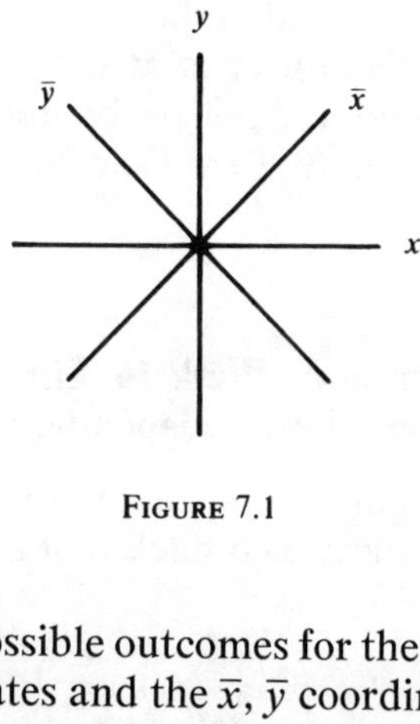

FIGURE 7.1

Look at the possible outcomes for the first step using the x, y coordinates and the $\bar{x}, \bar{y}$ coordinates. We have

x, y coordinates	$\bar{x}, \bar{y}$ coordinates
$(0, 1)$	$(1/\sqrt{2}, 1/\sqrt{2})$
$(0, -1)$	$(-1/\sqrt{2}, -1/\sqrt{2})$
$(1, 0)$	$(1/\sqrt{2}, -1/\sqrt{2})$
$(-1, 0)$	$(-1/\sqrt{2}, 1/\sqrt{2})$

Assume that we have two independent walkers, one moving with step size $1/\sqrt{2}$ randomly along the $\bar{x}$ axis and the other moving with the same step size along the $\bar{y}$ axis. Then, if we plot their positions using the $\bar{x}, \bar{y}$ axes, the possible outcomes for the first step would agree with those given for the above $\bar{x}, \bar{y}$ coordinates. The probabilities for each pair of outcomes would also be $((1/2)) \cdot ((1/2)) = 1/4$. Thus, we cannot distinguish a simple random walk in two dimensions from two independent walkers along the $\bar{x}$ and $\bar{y}$ axes making steps of magnitude $1/\sqrt{2}$. Since this probability of return does not depend upon the magnitude of the steps, the probability that our two independent walkers are at $(0, 0)$ at time $2n$ is equal to the product of the probabilities that each separately is at 0 at time $2n$, namely $(1/2^{2n})\binom{2n}{n}$. Therefore, the probability that the standard walk will be at $(0, 0)$ at time $2n$ is $\left((1/2^{2n})\binom{2n}{n}\right)^2$ as observed earlier.

7.7 Simple Random Walk in Three Dimensions is Not The Same as Three Independent Random Walks

In three dimensions, the probability that three independent walkers are each back to $\mathbf{0}$ after time $2n$ is

$$v_{2n} = \left(\binom{2n}{n}\frac{1}{2^{2n}}\right)^3.$$

This does not agree with our result for a simple random walk in three dimensions. Hence, the same trick cannot work. However, it is interesting to consider a random walk which is the result of three independent walkers. Let (i, j, k) be the position of three independent random walkers. The next position is one of the eight possibilities

$$(i+1, j+1, k+1),$$
$$(i+1, j+1, k-1),$$
$$(i+1, j-1, k+1),$$
$$(i+1, j-1, k-1),$$
$$(i-1, j+1, k+1),$$
$$(i-1, j+1, k-1),$$
$$(i-1, j-1, k+1),$$
$$(i-1, j-1, k-1).$$

Thus we may regard their progress as a random walk on the lattice points in three dimensions. If we center a cube of side 2 at (i, j, k), then the walk moves with equal probabilities to each of the eight corners of the cube. It is easier to show that this random walk is transient (using classical methods) than it is for simple random walk. This is because we can again use the one-dimension calculation. The probability v_{2n} for return at time $2n$ is

$$v_{2n} = \left(\binom{2n}{n}\frac{1}{2^{2n}}\right)^3 \sim \left(\frac{1}{\sqrt{\pi n}}\right)^3.$$

Thus

$$m = \sum_n v_{2n} \sim \sum_n \left(\frac{1}{\pi n}\right)^{3/2} < \infty$$

and the walk is transient.

The fact that this three independent walkers model and simple random walk are of the same type suggests that when two random walks are "really about the same," they should either both be transient or both be recurrent. In the next section we will see that this is indeed the case. Thus we may infer the transience of simple random walk in 3 dimensions from the transience of the three independent walkers model without going through the involved calculation of Section 7.4.

CHAPTER 8

RANDOM WALKS ON MORE GENERAL INFINITE NETWORKS

8.1 Random Walks on Infinite Networks

From now on we assume that G is an infinite connected graph. We assume that it is of **bounded degree**, by which we mean that there is some integer E such that the number of edges from any point is at most E. We assign to each edge xy of G a conductance $C_{xy} > 0$ with resistance $R_{xy} = 1/C_{xy}$. The graph G together with the conductances $\mathbf{C} = (C_{xy})$ is called a **network** and denoted by $(G, \mathbf{C})$. Given a network $(G, \mathbf{C})$, we define a random walk by

$$P_{xy} = \frac{C_{xy}}{C_x}$$

where $C_x = \Sigma_y C_{xy}$. When all the conductances are equal, we obtain a random walk that moves along each edge with the same probability: In agreement with our previous terminology, we call this walk **simple random walk** on G.

We have now a quite general class of infinite state Markov chains. As in the case of finite networks, the chains are **reversible Markov chains**: That is, there is a positive vector $\mathbf{w}$ such that $w_x P_{xy} = w_y P_{yx}$. As in the finite case, we can take $w_x = C_x$, since $C_x P_{xy} = C_{xy} = C_{yx} = C_y P_{yx}$.

8.2 The Type Problem

Let $(G, \mathbf{C})$ be an infinite network with random walk $\mathbf{P}$. Let $\mathbf{0}$ be a reference point. Let p_{escape} be the probability that a walk starting at $\mathbf{0}$ will never return to $\mathbf{0}$. If $p_{\text{escape}} = 0$ we say that $\mathbf{P}$ is **recurrent**, and if $p_{\text{escape}} > 0$ we say that it is **transient**. You are asked to show in Exercise 1 that the question of recurrence or transience of $\mathbf{P}$ does not depend upon the choice of the reference point. The **type problem** is the problem of determining if a random walk (network) is recurrent or transient.

In Section 5.5 we showed how to rephrase the type problem for a lattice in terms of finite graphs sitting inside it. In Section 5.6 we showed that the type problem is equivalent to an electrical network problem by showing that simple random walk on a lattice is recurrent if and only if the lattice has infinite resistance to infinity. The same arguments apply with only minor modifications to the more general infinite networks as well. This means that we can use Rayleigh's short-cut method to determine the type of these more general networks.

Exercise

1. Show that the question of recurrence or transience of $\mathbf{P}$ does not depend upon the choice of the reference point.

8.3 Comparing Two Networks

Given two sets of conductances $\mathbf{C}$ and $\bar{\mathbf{C}}$ on G, we say that $(G, \bar{\mathbf{C}}) \leqslant (G, \mathbf{C})$ if $\bar{C}_{xy} \leqslant C_{xy}$ for all xy, or equivalently, if $\bar{R}_{xy} \geqslant R_{xy}$ for all xy. Assume that $(G, \bar{\mathbf{C}}) \leqslant (G, \mathbf{C})$. Then by the Monotonicity Law, $\bar{R}_{EFF} \geqslant R_{EFF}$. Thus if random walk on $(G, \bar{\mathbf{C}})$ is transient, i.e., if $\bar{R}_{EFF} < \infty$, then random walk on $(G, \mathbf{C})$ is also transient. If random walk on $(G, \mathbf{C})$ is recurrent, i.e., if $R_{EFF} = \infty$, then random walk on $(G, \bar{\mathbf{C}})$ is also recurrent.

THEOREM. *If $(G, \mathbf{C})$ and $(G, \mathbf{C}')$ are networks, and if there exist constants u, v with $0 < u \leqslant v < \infty$ such that*

$$uC_{xy} \leqslant C'_{xy} \leqslant vC_{xy}$$

for all x and y, then random walk on $(G, \mathbf{C}')$ is of the same type as random walk on $(G, \mathbf{C})$.

Proof. Let $U_{x,y} = uC_{x,y}$ and $V_{x,y} = vC_{x,y}$. Then $(G, \mathbf{U}) \leqslant (G, \mathbf{C}') \leqslant (G, \mathbf{V})$. But the random walks for $(G, \mathbf{U})$ and $(G, \mathbf{V})$ are the same as random walk on $(G, \mathbf{C})$. Thus random walk for $(G, \mathbf{C}')$ is of the same type as random walk on $(G, \mathbf{C})$.

COROLLARY. *Let $(G, \mathbf{C})$ be a network. If for every edge xy of G we have $0 < u \leqslant C_{xy} \leqslant v < \infty$ for some constants u and v, then the random walk on $(G, \mathbf{C})$ has the same type as simple random walk on G.*

EXERCISES

1. Consider the two-dimensional lattice. For each edge, we toss a coin to decide what kind of resistor to put across this edge. If heads turns up, we put a two-ohm resistor across this edge; if tails turns up, we put a one-ohm resistor across the edge. Show that the random walk on the resulting network is recurrent.

2. Consider the analogous problem in 3 dimensions.

8.4 The k-Fuzz of a Graph

For any integer k, the k-**fuzz** of a graph G is the graph G_k obtained from G by adding an edge xy if it is possible to go from x to y in at most k steps. For example, the 2-fuzz of the two-dimensional lattice is shown in Figure 8.1; please note that horizontal and vertical edges of length 2, such as those joining (0, 0) to (0, 2), have not been indicated.

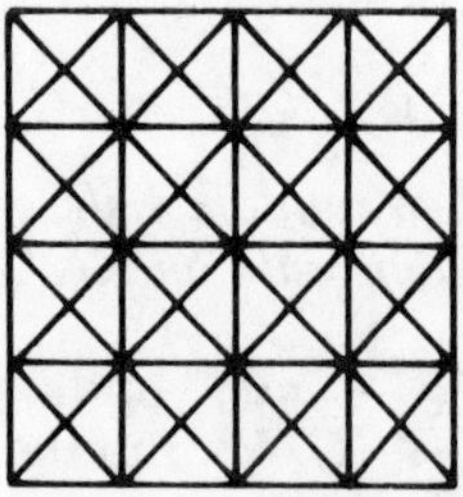

FIGURE 8.1

THEOREM. *Simple random walk on G and on the k-fuzz G_k of G have the same type.*

Proof. Let $\mathbf{P}$ be simple random walk on G. Define $\bar{\mathbf{P}} = (\mathbf{P} + \mathbf{P}^2 + \cdots + \mathbf{P}^k)/k$. Then $\bar{\mathbf{P}}$ may be considered to be $\mathbf{P}$, watched at one of the first k steps chosen at random, then at a time chosen at random from the next k steps after this time, etc. Thinking of $\bar{\mathbf{P}}$ in this way, we see that $\mathbf{P}$ is in state $\mathbf{0}$ at least once for every time $\bar{\mathbf{P}}$ in state $\mathbf{0}$. Hence, if $\bar{\mathbf{P}}$ is recurrent so is $\mathbf{P}$. Assume now that $\bar{\mathbf{P}}$ is transient. Choose a finite set E so that $\mathbf{0}$ cannot be reached in k steps from a point outside of E. Then, since the walk $\bar{\mathbf{P}}$ will be outside E from some time on, the walk $\mathbf{P}$ cannot be at $\mathbf{0}$ after this time, and $\mathbf{P}$ is also transient. Therefore, $\mathbf{P}$ and $\bar{\mathbf{P}}$ are of the same type.

Finally, we show that $\bar{\mathbf{P}}$ has the same type as simple random walk on G_k. Here it is important to remember our restriction that G is of bounded degree, so that for some E no vertex has degree $>E$. We know that $\mathbf{P}$ is reversible with $\mathbf{wP} = \mathbf{w}$ where w_x is the number of edges coming out of x. From its construction, $\bar{\mathbf{P}}$ is also reversible and $\mathbf{w}\bar{\mathbf{P}} = \mathbf{w}$. $\bar{\mathbf{P}}$ is the random walk on a network $(G_k, \bar{\mathbf{C}})$ with $\bar{C}_{xy} = w_x \bar{P}_{xy}$. If $\bar{P}_{xy} > 0$, there is a path $x, x_1, x_2, \ldots, x_{m-1}, y$ in G from x to y of length $m \leqslant k$. Then $\bar{P}_{xy} \geqslant (1/k)(1/E)^m \geqslant (1/k)(1/E)^k$. Thus $0 < (1/k)E^k \leqslant \bar{P}_{xy} \leqslant 1$ and $(1/k)E^k \leqslant \bar{C}_{xy} \leqslant E$. Therefore, by the theorem on the irrelevance of bounded twiddling proven in Section 8.3, $\bar{\mathbf{P}}$ and simple random walk on G_k are of the same type. So G and G_k are of the same type.

NOTE: This is the only place where we use probabilistic methods of proof. For the purist who wishes to avoid probabilistic methods, Exercise 2 of Section 8.6 indicates an alternative electrical proof.

We show how this theorem can be used. We say that a graph G can be **embedded** in a graph $\bar{G}$ if the points x of G can be made to correspond in a one-to-one fashion to points $\bar{x}$ of $\bar{G}$ in such a way that if xy is an edge in G, then $\bar{x}\bar{y}$ is an edge in $\bar{G}$.

THEOREM. *If simple random walk on G is transient, and if G can be embedded in a k-fuzz $\bar{G}_k$ of $\bar{G}$, then simple random walk on $\bar{G}$ is also transient. Simple random walk on G and $\bar{G}$ are of the same type if each graph can be embedded in a k-fuzz of the other graph.*

Proof. Assume that simple random walk on G is transient and that G can be embedded in a k-fuzz $\bar{G}_k$ of $\bar{G}$. Since R_{EFF} for G is finite and G can be embedded in $\bar{G}_k$, R_{EFF} for $\bar{G}_k$ is finite. By our previous theorem, the same is true for $\bar{G}$ and simple random walk on $\bar{G}$ is transient.

If we can embed G in $\bar{G}_k$ and $\bar{G}$ in G_k, then the random walk on G is transient if and only if the random walk on $\bar{G}$ is.

EXERCISE

1. We have assumed that there is a bound E for the number of edges coming out of any point. Show that if we do not assume this, it is not necessarily true that G and G_k are of the same type. (Hint: Consider a network something like that shown in Figure 8.2.)

FIGURE 8.2

8.5 Comparing General Graphs with Lattice Graphs

We know the type of simple random walk on a lattice $\mathbf{Z}^d$. Thus to determine the type of simple random walk on an arbitrary graph G, it is natural to try to compare G with $\mathbf{Z}^d$. This is feasible for graphs that can be drawn in some Euclidean space $\mathbf{R}^d$ in a civilized manner.

DEFINITION. A graph G can be drawn in a Euclidean space $\mathbf{R}^d$ in a **civilized manner** if its vertices can be embedded in $\mathbf{R}^d$ so that for some $r < \infty$, $s > 0$,

(a) The length of each edge is $\leqslant r$.

(b) The distance between any two points is $\geqslant s$.

Note that we make no requirement about being able to draw the edges of G so they don't intersect.

THEOREM. *If a graph can be drawn in* $\mathbf{R}^d$ *in a civilized manner, then it can be embedded in a k-fuzz of the lattice* $\mathbf{Z}^d$.

Proof. We carry out the proof for the case $d = 2$. Assume that G can be drawn in a civilized manner in $\mathbf{R}^2$. We want to show that G can be embedded in a k-fuzz of $\mathbf{Z}^2$. We have been thinking of $\mathbf{Z}^2$ as being drawn in $\mathbf{R}^2$ with perpendicular lines and adjacent points a unit distance apart on these lines, but this embedding is only one particular way of representing $\mathbf{Z}^2$. To emphasize this, let's talk about L^2 instead of $\mathbf{Z}^2$. Figure 8.3 shows another way of drawing L^2 in R^2. From a graph-theoretical point of view, this is the same as $\mathbf{Z}^2$. In trying to compare G to L^2, we take advantage of this flexibility by drawing L^2 so small that points of G can be moved onto points of L^2 without bumping into each other.

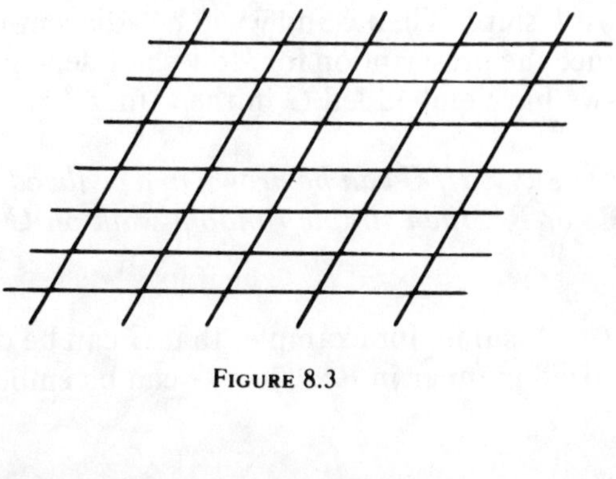

FIGURE 8.3

Specifically, let L^2 be a two-dimensional rectangular lattice with lines a distance $s/2$ apart. In any square of L^2, there is at most one point of G. Move each point $\mathbf{x}$ of $\mathbf{G}$ to the southwest corner $\bar{x}$ of the square that it is in, as illustrated in Figure 8.4.

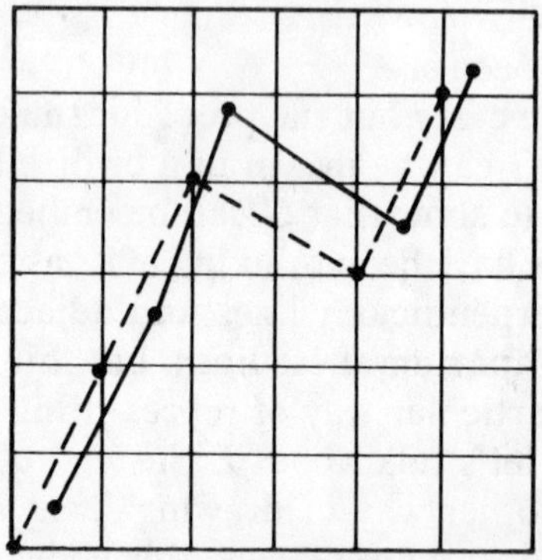

FIGURE 8.4

Now since any two adjacent points x, y in G were within r of each other in $\mathbf{R}^2$, the corresponding points $\bar{x}$, $\bar{y}$ in L^2 will have Euclidean distance $\leqslant r + 2s$. Choose k so that any two points of L^2 whose Euclidean distance is $\leqslant r + 2s$ can be connected by a path in L^2 of at most k steps. Then $\bar{x}$ and $\bar{y}$ will be adjacent in L_k^2, and—since the prescription for k does not depend on x and y—we have embedded G in the k-fuzz L_k^2.

COROLLARY. *If G can be drawn in a civilized manner in* $\mathbf{R}^1$ *or* $\mathbf{R}^2$, *then simple random walk on G is recurrent.*

Proof. Assume, for example, that G can be drawn in a civilized manner in $\mathbf{R}^2$. Then G can be embedded in a k-fuzz $\mathbf{Z}_k^2$ of $\mathbf{Z}^2$. If simple random walk on G were transient, then the same would be true for $\mathbf{Z}_k^2$ and $\mathbf{Z}^2$. But we know that simple random walk on $\mathbf{Z}^2$ is recurrent. Thus simple random walk on G is recurrent.

Our first proof that random walk in three dimensions is transient consisted in showing that we could embed a transient tree in $\mathbf{Z}^3$. We now know that it would have been sufficient to show how to draw a transient tree in $\mathbf{R}^3$ in a civilized manner: This is easier (see Exercise 1).

The corollary implies that simple random walk on any sufficiently symmetrical graph in $\mathbf{R}^2$ is recurrent. For example, simple random walk on the regular graph made up of hexagons shown in Figure 8.5 is recurrent.

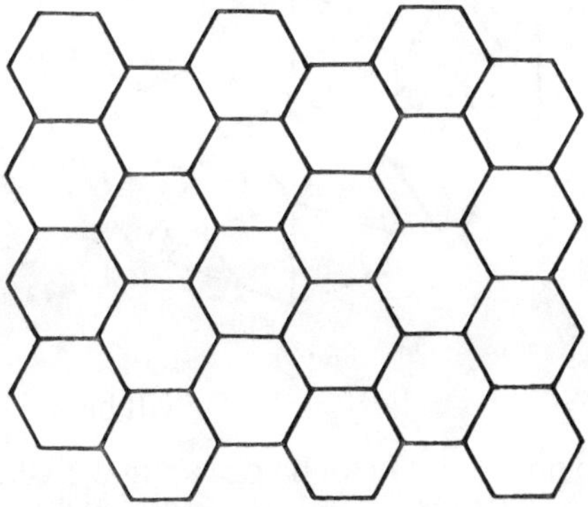

FIGURE 8.5

We can even consider very irregular graphs. For example, on the cover of the January 1977 Scientific American, there is an example due to Conway of an infinite non-periodic tiling using Penrose tiles of the form shown in Figure 8.6. It is called the cartwheel pattern; part of it is shown in Figure 8.7. A walker walking randomly on the edges of this very irregular infinite tiling will still return to his starting point.

FIGURE 8.6

FIGURE 8.7

Assume now that G can be drawn in a civilized manner in $\mathbf{R}^3$. Then to show that simple random walk on G is of the same type as $\mathbf{Z}^3$, namely transient, it is sufficient to show that we can embed $\mathbf{Z}^3$ in a k-fuzz of G. This is clearly possible for any regular lattice in $\mathbf{R}^3$. The three lattices that have been most studied and for which exact probabilities for return have been found are called the SC, BCC, and FCC lattices. The SC (simple cubic) lattice is just $\mathbf{Z}^3$. The walker moves each time to a new point by adding a random choice from the six vectors $(\pm 1, 0, 0)$, $(0, \pm 1, 0)$, $(0, 0, \pm 1)$. For the BCC (body centered cubic) lattice, the choice is one of the eight vectors $(\pm 1, \pm 1, \pm 1)$. This was the walk that resulted from three independent one-dimensional walkers. For the FCC (face centered cubic) lattice, the random choice is made from the twelve vectors $(\pm 1, \pm 1, 0)$, $(\pm 1, 0, \pm 1)$, $(0, \pm 1, \pm 1)$. For a discussion of exact calculations for these three lattices, see Montroll and West [**25**].

As we have seen, once the transience of any one of these three walks is established, no calculations are necessary to determine that the other walks are transient also. Thus we have yet another way of establishing Pólya's theorem in three dimensions: Simply verify transience of the walk on the BCC lattice via the simple three independent walkers computation, and infer that walk on the SC lattice is also transient since the BCC lattice can be embedded in a k-fuzz of it.

Exercises

1. When we first set out to prove Pólya's theorem for $d = 3$ way back in Section 3.2, our idea was to embed NT_3 in $\mathbf{Z}^3$. As it turned out, what we ended up embedding was not NT_3 but $NT_{2.5849\ldots}$, and we didn't quite embed it at that. We tried to improve the situation by finding (in Exercise 3 of Section 6.9) an honest-to-goodness embedding of a relative of $NT_{2.5849\ldots}$, but NT_3 was still left completely out in the cold. Now, however, we are in a position to embed NT_3, if not in $\mathbf{Z}^3$ then at least in a k-fuzz of it. All we need to do is to draw NT_3 in $\mathbf{R}^3$ in a civilized manner. Described how to do this, and thereby give one more proof of Pólya's theorem for $d = 3$.

2. Find a graph that can be embedded in a civilized manner in $\mathbf{R}^3$ but not in $\mathbf{R}^2$, but is nonetheless recurrent.

3. Assume that G is drawn in a civilized manner in $\mathbf{R}^3$. To show that simple random walk on G is transient, it is enough to know that $\mathbf{Z}^3$ can be embedded in a k-fuzz of G. Try to come up with a nice condition that will guarantee that this is possible. Can you make this condition simple, yet general enough so that it will settle all reasonably interesting cases? In other words, can you make the condition nice enough to allow us to remember only the condition, and forget about the general method lying behind it?

8.6 Solving the Type Problem by Flows—a Variant of the Cutting Method

In this section we will introduce a variant of the cutting method whereby we use Thomson's Principle directly to estimate the effective resistance of a conductor.

Thomson's Principle says that, given any unit flow through a resistive medium, the dissipation rate of that flow gives an upper bound for the effective resistance of the medium. This suggests that to show that a given infinite network is transient, it should be enough to produce a unit flow out to infinity having finite energy dissipation.

In analogy with the finite case, we say that $\mathbf{j}$ is a **flow from 0 to infinity** if

(a) $j_{xy} = -j_{yx}$

(b) $\sum_y j_{xy} = 0 \qquad x \neq 0.$

We define $j_0 = \Sigma_y j_{0y}$. If $j_0 = 1$, we say that $\mathbf{j}$ is a **unit flow** to infinity. Again in analogy with the finite case, we call $(1/2)\ \Sigma_{x,y}\, j_{xy}^2 R_{xy}$ the **energy dissipation** of the flow j_{xy}.

THEOREM. *The effective resistance* R_{EFF} *from* **0** *to infinity is less than the energy dissipation of any other unit flow from* **0** *to infinity.*

Proof. Assume that we have a unit flow $\mathbf{j}$ from **0** to infinity with energy dissipation $E = (1/2)\ \Sigma_{x,y} j_{xy}^2 R_{xy}$. We claim that $R_{EFF} \leqslant E$. Restricting j_{xy} to the edges of the finite graph $G^{(r)}$, we have a unit flow from **0** to $S^{(r)}$ in $G^{(r)}$. Let $\mathbf{i}^{(r)}$ be the unit current flow in $G^{(r)}$ from **0** to $S^{(r)}$. By the results of Section 3.5 of Part I,

$$
\begin{aligned}
R_{EFF}^{(r)} &= \frac{1}{2} \sum_{G^{(r)}} (i_{xy}^{(r)})^2 R_{xy} \\
&\leqslant \frac{1}{2} \sum_{G^{(r)}} j_{xy}^2 R_{xy} \\
&\leqslant \frac{1}{2} \sum_{x,y} j_{xy}^2 R_{xy} = E,
\end{aligned}
$$

where $\Sigma_{G^{(r)}}$ is short for

$$
\sum_{\substack{x,y \\ xy \text{ an edge of } G^{(r)}}} .
$$

EXERCISES

1. We have billed the method of using Thomson's Principle directly to estimate the effective resistances of a network as a variant of the cutting method. Since the cutting method was derived from Thomson's Principle, and not vice versa, it would seem that we have got the cart before the horse. Set this straight by giving an informal ("heuristic") derivation of Thomson's Principle from the cutting method. (Hint: see Maxwell [**23**], Chapter VIII, Paragraph 307.) For more on this question, see Onsager [**27**].

2. Let G be an infinite network and G_k the network obtained from the k-fuzz of G. Using electric network argu-

ments, show that $R_{EFF} < \infty$ for G if and only if $R_{EFF} < \infty$ for G_k.

8.7 A Proof, Using Flows, that Simple Random Walk in Three Dimensions is Transient

We now apply this form of the cutting method to give another proof that simple random walk on the three-dimensional lattice is transient. All we need is a flow to infinity with finite dissipation. The flow we are going to describe is not the first flow one would think of. In case you are curious, the flow described here was constructed as a side effect of an unsuccessful attempt to derive the isoperimetric inequality (see Pólya [**29**]) from the "max-flow min-cut" theorem (Ford and Fulkerson [**7**]). The idea is to find a flow in the positive orthant having the property that the same amount flows through all points at the same distance from **0**.

Again, it is easiest to show the construction for the two-dimensional case. Let G denote the part of $\mathbf{Z}^2$ lying in the first quadrant. The graph $G^{(4)}$ is shown in Figure 8.8.

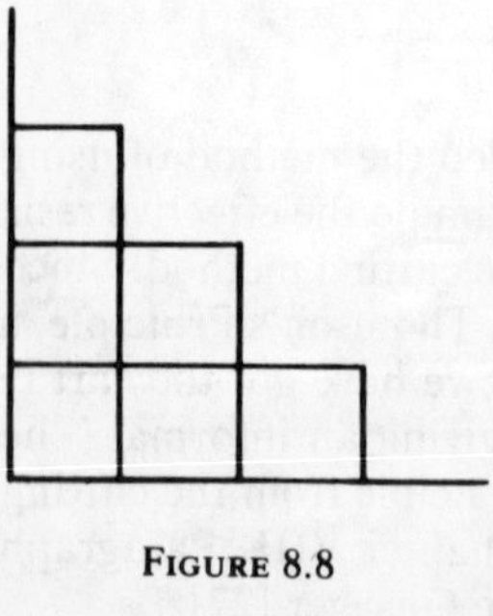

FIGURE 8.8

We choose our flow so that it always goes away from **0**. Into each point that is not on either axis there are two flows, one vertical and one horizontal. We want the sum of the corresponding values of j_{xy} to be the same for all points the same distance from $\mathbf{0}$. These conditions completely determine the flow. The flow out of the point x, y with $x + y = n$ is as shown in Figure 8.9. The values for the currents out to the fourth level are shown in Figure 8.10.

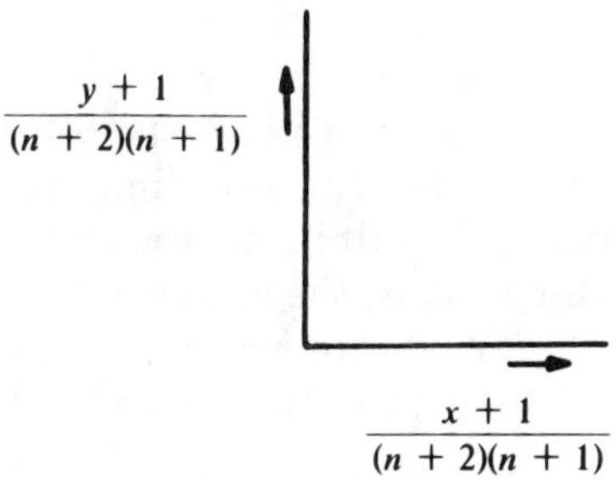

FIGURE 8.9

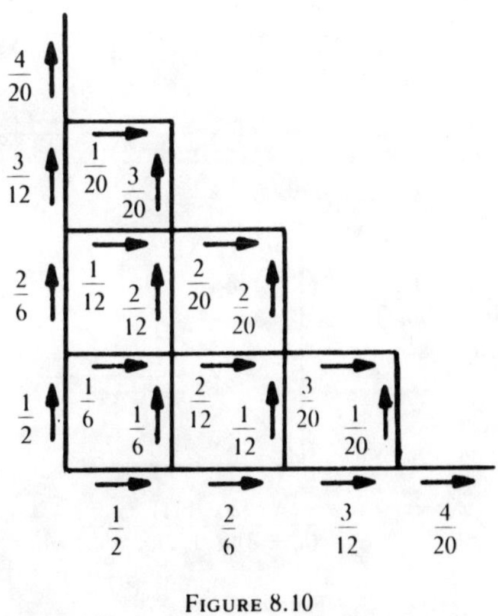

FIGURE 8.10

In general, the flow out of a point (x, y) with $x + y = n$ is

$$\frac{x+1}{(n+2)(n+1)} + \frac{y+1}{(n+2)(n+1)} = \frac{1}{n+1}$$

and the flow into this point is

$$\frac{x}{n(n+1)} + \frac{y}{n(n+1)} = \frac{1}{n+1}.$$

Thus the net flow at (x, y) is 0. The flow out of $\mathbf{0}$ is $(1/2) + (1/2) = 1$. For this two-dimensional flow, the energy dissipation is infinite as it would have to be.

For three dimensions, the uniform flow is defined as follows: out of (x, y, z) with $x + y + z = n$ we have the flow indicated in Figure 8.11. The total flow out of (x, y, z) is then

$$\frac{2(x+1)}{(n+3)(n+2)(n+1)} + \frac{2(y+1)}{(n+3)(n+2)(n+1)} + \frac{2(z+1)}{(n+3)(n+2)(n+1)} = \frac{2}{(n+2)(n+1)}.$$

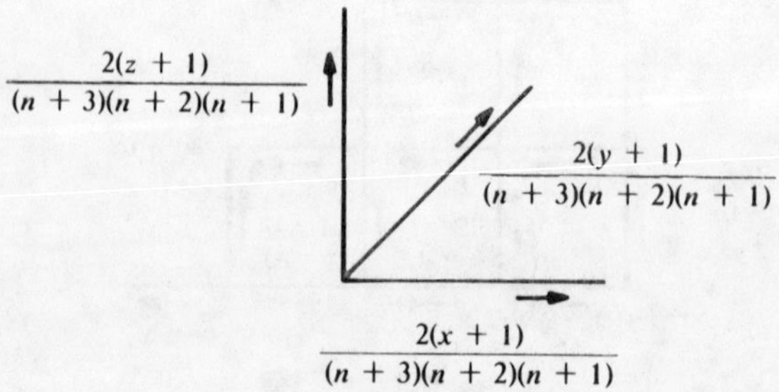

FIGURE 8.11

The flow into (x, y, z) comes from the points $(x-1, y, z)$, $(x, y-1, z)$, $(x, y, z-1)$ and, hence, the total flow into (x, y, z) is

$$\frac{2x}{(n+2)(n+1)n} + \frac{2y}{(n+2)(n+1)n} + \frac{2z}{(n+2)(n+1)n}$$

$$= \frac{2}{(n+2)(n+1)}.$$

Thus the net flow for (x, y, z) is 0. The flow out of $\mathbf{0}$ is $(1/3) + (1/3) + (1/3) = 1$. We have now to check finiteness of energy dissipation. The flows coming out of the edges at the nth level are all $\leqslant 2/(n+1)^2$. There are $(n+1)(n+2)/2$ points a distance n from 0, and thus there are $(3/2)(n+1)(n+2) \leqslant 3(n+1)^2$ edges coming out of the nth level. Thus the energy dissipation E has

$$E \leqslant \sum_n 3(n+1)^2 \left(\frac{2}{(n+1)^2}\right)^2 = 12 \sum_n \frac{1}{(n+1)^2} < \infty$$

and the random walk is transient.

8.8 The End

We have come to the end of our labors, and it seems fitting to look back and try to say what it is we have learned.

To begin with, we have seen how phrasing certain mathematical questions in physical terms allows us to draw on a large body of physical lore, in the form of established methods and ways of thought, and thereby often leads us to the answers to those questions.

In particular, we have seen the utility of considerations involving energy. In took hundreds of years for the concept of energy to emerge and take its rightful place in physical theory, but it is now recognized as perhaps the most fundamental concept in all of physics. By phrasing our probabilistic problems in physical terms, we were naturally led to considerations of energy, and these considerations showed us the way through the difficulties of our problems.

As for Pólya's theorem and the type problem in general, we have picked up a bag of tricks, known collectively as "Rayleigh's short-cut method," which we may expect will allow us to determine the type of almost any random walk we are likely to embark on. In the process, we have gotten some feeling for the connection between the dimensionality of a random walk and its type. Furthermore, we have settled one of the main questions likely to occur to someone encountering Pólya's theorem, namely: "If two walks look essentially the same, and if one has been shown to be transient, must not the other also be transient?"

Another question likely to occur to someone contemplating Pólya's theorem is the question raised in Section 5.8: "Since the lattice $\mathbf{Z}^d$ is in some sense a discrete analog of a resistive medium filling all of $\mathbf{R}^d$, should it not be possible to go quickly and naturally from the trivial computation of the resistance to infinity of the continuous medium to a proof of Pólya's theorem?" Our shorting argument allowed us to do this in the two-dimensional case; that leaves the case of three (or more) dimensions. Again, it is considerations of energy that allow us to make this connection. The trick is to start with the flow field that one gets by solving the continuous problem, and adapt it to the lattice, so as to get a lattice flow to infinity having finite dissipation rate. We leave the working out of this as an exercise, so as not to rob readers of the fun of doing it for themselves.

Exercise

1. Give one final proof of Pólya's theorem in 3 dimensions by showing how to adapt the $1/r^2$ radial flow field to the lattice. (Hint: "cubes.")

REFERENCES

1. Abbott, Edwin, *Flatland*, Little, Brown and Company, Boston, 1899.

2. Bollobás, B., *Graph Theory: An Introductory Course*, Springer-Verlag, New York, 1979.

3. Courant, R., K. Friedrichs, and H. Lewy, Über die partiellen Differenzengleichungen der mathematischen Physik, *Math. Ann.*, 100 (1928), 32–74. Translation: On the Partial Difference Equations of Mathematical Physics, *IBM J. Research and Development*, 11 (1967), 215–237.

4. Doob, J. L., *Stochastic Processes*, John Wiley & Sons, New York, 1953.

5. Feller, William, *An Introduction to Probability Theory and Its Applications*, Vol I, 3rd edition, John Wiley & Sons, New York, 1968.

6. Feynman, Richard, *The Feynman Lectures on Physics*, Addison-Wesley, Reading, Mass, 1964.

7. Ford, L. R. and D. R. Fulkerson, Maximal Flow through a Network, *Can J. Math.*, 8 (1956), 399–404.

8. Freedman, D., R. Pisani, and R. Purves, *Statistics*, W. W. Norton, New York, 1978.

9. Glasser, M. L. and I. J. Zucker, Extended Watson Integrals for the Cubic Lattices, *Proc. Natl. Acad. Sci.*, USA, 74 (1977), 1800–1801.

10. Griffeath, D. and T. M. Liggett, Critical Phenomena for Spitzer's Reversible Nearest Particle Systems, *Ann. of Probability*, 10 (1982), 881–895.

11. Hersh, R. and R. J. Griego, Brownian Motion and Potential Theory, *Scientific American* (March 1969), 66–74.

12. Jeans, James, *The Mathematical Theory of Electricity and Magnetism*, 5th Edition, Cambridge, 1966.

13. Kakutani, S., Markov Processes and the Dirichlet Problem, *Proc. Jap. Acad.*, 21 (1945), 227–233.

14. Kelly, F., *Reversibility and Stochastic Networks*, John Wiley & Sons, New York, 1979.

15. Kemeny, J. G., J. L. Snell, and A. W. Knapp, *Denumberable Markov Chains*, Springer-Verlag, New York, 1976.

16. Kemeny, J. G., J. L. Snell, and G. L. Thompson, *Finite Mathematics*, 3rd Edition, Prentice-Hall, Englewood Cliffs, N.J., 1974.

17. Kesten, Harry, *Percolation Theory for Mathematicians*, Birkhäuser, Boston, 1982.

18. Kingman, J. F. C., Markov Population Processes, *J. Appl. Prob.*, 6 (1969), 1–18.

19. Kirchhoff, Gustav, *Vorlesungen über Electricität und Magnetismus*, Leipzig, 1891.

20. Lehman, A., A Resistor Network Inequality, Problem 60-5, *SIAM Review*, 4 (1965), 150–154.

21. Lévy, Paul, *Théorie de l'addition des variables aléatoires*, Paris, 1937.

22. Lyons, T., A Simple Criterion for Transience of a Reversible Markov Chain, *Ann. of Probability*, 11 (1983), 393–402.

23. Maxwell, James Clerk, *A Treatise on Electricity and Magnetism*, 3rd Edition, Clarendon, 1918, reprinted by Dover, New York, 1954.

24. McCrea, W. H. and F. J. W. Whipple, Random Paths in Two and Three Dimensions, *Proc. of the Royal Soc. of Edinburgh*, 60 (1940), 281–298.

25. Montroll, E. W. and B. J. West, Fluctuation Phenomena, in: Lebowitz and Montroll, eds., *Studies in Statistical Mechanics*, 7, North-Holland, 1979, pp. 61–175.

26. Nash-Williams, C. St. J. A., Random Walks and Electric Currents in Networks, *Proc. of the Cambridge Philosophical Soc.*, 55 (1959), 181–194.

27. Onsager, L., Reciprocal Relations in Irreversible Processes, *Phys. Rev.*, 37 (1931), 405–426; 38 (1931), 2265–2279.

28. Pólya, G. Über eine Aufgabe der Wahrscheinlichkeitsrechnung betreffend die Irrfahrt im Strassennetz, *Mathematische Annalen*, 84 (1921), 149–160.

29. Pólya, G., *How To Solve It*, 2nd Edition, Princeton, 1957.

30. Pólya, G. and G. Szegö, *Isoperimetric Inequalities of Mathematical Physics*, Princeton, 1951.

31. Rayleigh, J. W. S., On the Theory of Resonance, *Philosophical Transactions*, CLXI (1870), reprinted in: *Collected Scientific Papers*, Vol 1, Cambridge, 1899, pp. 33–75.

32. Royden, H. L., Harmonic Functions on Open Riemann Surfaces, *Trans. Amer. Math. Soc.*, 73 (1952).

33. Shannon, C. E. and D. W. Hagelbarger, Concavity of Resistance Functions, *J. of Appl. Phys.*, 27 (1956), 42–43.

34. Snell, J. L., Gambling, Probability and Martingales, *The Mathematical Intelligencer*, Vol. 4, No. 3, 1982.

35. Thomson, W. and P. G. Tait, *Treatise on Natural Philosophy*, Cambridge, 1879.

36. Ville, J., *Étude critiques de la notion de collectif*, Gauthier-Villars, Paris, 1939.

37. Watson, G. N., Three Triple Integrals, *Quarterly J. Math.* (Oxford), 10 (1939), 266–276.

INDEX